Hydrogeologic and Geochemical Characterization of Groundwater Resources in Rush Valley, Tooele County, Utah

By Philip M. Gardner and Stefan Kirby

Prepared in cooperation with the State of Utah Department of Natural Resources

Scientific Investigations Report 2011–5068

U.S. Department of the Interior
U.S. Geological Survey

U.S. Department of the Interior
KEN SALAZAR, Secretary

U.S. Geological Survey
Marcia K. McNutt, Director

U.S. Geological Survey, Reston, Virginia: 2011

For more information on the USGS—the Federal source for science about the Earth, its natural and living resources, natural hazards, and the environment, visit http://www.usgs.gov or call 1-888-ASK-USGS

For an overview of USGS information products, including maps, imagery, and publications, visit http://www.usgs.gov/pubprod

To order this and other USGS information products, visit http://store.usgs.gov

Suggested citation:
Gardner, P.M., and Kirby, S.M., 2011, Hydrogeologic and geochemical characterization of groundwater resources in Rush Valley, Tooele County, Utah: U.S. Geological Survey Scientific Investigations Report 2011–5068, 68 p.

Acknowledgements

The USGS acknowledges the support and cooperation of the municipalities and private land owners in Rush Valley that granted access to wetland areas, springs, and wells. Their time and interest and discussions of their respective water works and water-use practices are greatly appreciated.

Contents

Figures

Tables

Appendix 1

Conversion Factors

Multiply	By	To obtain
Length		
inch (in.)	2.54	centimeter (cm)
foot (ft)	0.3048	meter (m)
meter (m)	3.281	foot (ft)
mile (mi)	1.609	kilometer (km)
Area		
acre	0.004047	square kilometer (km^2)
square mile (mi^2)	2.590	square kilometer (km^2)
Volume		
liter (L)	1.057	quart
acre-foot (acre-ft)	1,233	cubic meter (m^3)
Flow rate		
acre-foot per year (acre-ft/yr)	1,233	cubic meter per year (m^3/yr)
Pressure		
millimeters of mercury (mmHg)	0.001316	atmosphere, standard (atm)
Specific capacity		
gallon per minute per foot (gal/min/ft)	0.2070	liter per second per meter [(L/s)/m]
Hydraulic conductivity		
foot per day (ft/d)	0.3048	meter per day (m/d)
Transmissivity*		
foot squared per day (ft^2/d)	0.09290	meter squared per day (m^2/d)

Temperature in degrees Celsius (°C) may be converted to degrees Fahrenheit (°F) as follows: °F=(1.8×°C)+32

Vertical coordinate information is referenced to the National Geodetic Vertical Datum of 1929 (NGVD 29).

Horizontal coordinate information is referenced to the North American Datum of 1983 (NAD 83).

Altitude, as used in this report, refers to distance above the vertical datum.

*Transmissivity: The standard unit for transmissivity is cubic foot per day per square foot times foot of aquifer thickness [(ft^3/d)/ft^2]ft. In this report, the mathematically reduced form, foot squared per day (ft^2/d), is used for convenience.

Concentration of chemical constituents in water is reported in milligrams per liter (mg/L), micrograms per liter (μg/L), and in milliequivalents per liter. Milligrams per liter and micrograms per liter are units expressing the concentration of chemical constituents in solution as weight (grams) of solute per unit volume (liter) of water. A liter of water is assumed to weigh 1 kilogram, except for brines or water at high temperatures because of changes in the density of the water. For concentrations less than 7,000 mg/L or 7,000,000 μg/L, the numerical value is the same as for concentrations in parts per million or parts per billion, respectively. Milliequivalents per liter are units expressing concentrations that are chemically equivalent in terms of atomic or molecular weight and electrical charge.

Specific conductance is given in microsiemens per centimeter at 25 degrees Celsius (μS/cm at 25 °C). Concentrations of dissolved gases are reported in cubic centimeters of gas at standard temperature and pressure per gram of water (ccSTP/g). Tritium concentration is reported in tritium units (TU). The ratio of 1 atom of tritium to 1018 atoms of hydrogen is equal to 1 TU or 3.2 picocuries per liter. Carbon-14 activity is reported as percent modern carbon (pmc). Stable-isotope ratios are reported as delta (d) values, which are parts per thousand or permil (‰) differences from a standard.

Abbreviations and Acronyms

BCM	Basin Characterization Model.
BP	before present.
DCD	Deseret Chemical Depot.
EPA	Environmental Protection Agency.
GMWL	Global Meteoric Water Line.
HUC	Hydrologic Unit Code.
LBFAU	lower basin-fill aquifer unit.
LCAU	lower carbonate aquifer unit.
LMWL	local meteoric water line.
MCL	maximum contaminant level.
NAIP	National Agricultural Imagery Program.
NCCU	noncarbonate confining unit.
NOSAMS	National Ocean Sciences Accelerator Mass Spectrometry Facility.
NWIS	National Water Information System.
PRISM	Parameter-elevation Regressions on Independent Slopes Model.
SWReGAP	Southwest Regional Gap Analysis Project.
UBFAU	upper basin-fill aquifer unit.
UCAU	upper carbonate aquifer unit.
UGS	Utah Geological Survey.
UMWL	Utah meteoric water line.
USCU	upper siliciclastic confining unit.
USGS	U.S. Geological Survey.
VSMOW	Vienna Standard Mean Ocean Water.
VU	volcanic unit.
WRLU	water-related land-use survey.

Numbering system for hydrologic-data sites in Utah

The system of numbering wells, springs, and other hydrologic-data sites in Utah is based on the cadastral land-survey system of the U.S. Government. The number, in addition to designating the site, describes its position in the land net. The land-survey system divides the State of Utah into four quadrants by the Salt Lake Base Line and the Salt Lake Meridian—and in the Uinta Basin, by the Uintah Base Line and the Uintah Meridian. These quadrants are designated by the uppercase letters A, B, C, and D, which indicate, respectively, the northeast, northwest, southwest, and southeast quadrants. Numbers that designate the township and range, in that order, follow the quadrant letter, and all three are enclosed in parentheses. The number after the parentheses indicates the section and is followed by three lowercase letters that indicate the quarter section, the quarter-quarter section, and the quarter-quarter-quarter section—generally 10 acres for a regular section[1]. The lowercase letters a, b, c, and d indicate, respectively, the northeast, northwest, southwest, and southeast quarters of each subdivision. The number after the letters is the serial number of the well or spring within the 10-acre tract. Thus, (C-6-6)11ccc-1 desig-nates the first well visited in the SW ¼ of the SW ¼ of the SW ¼ of Sec. 11, T 6 S, R 6 W. The capital letter "C" indicates that the township is south of the Salt Lake Base Line and the range is west of the Salt Lake Meridian.

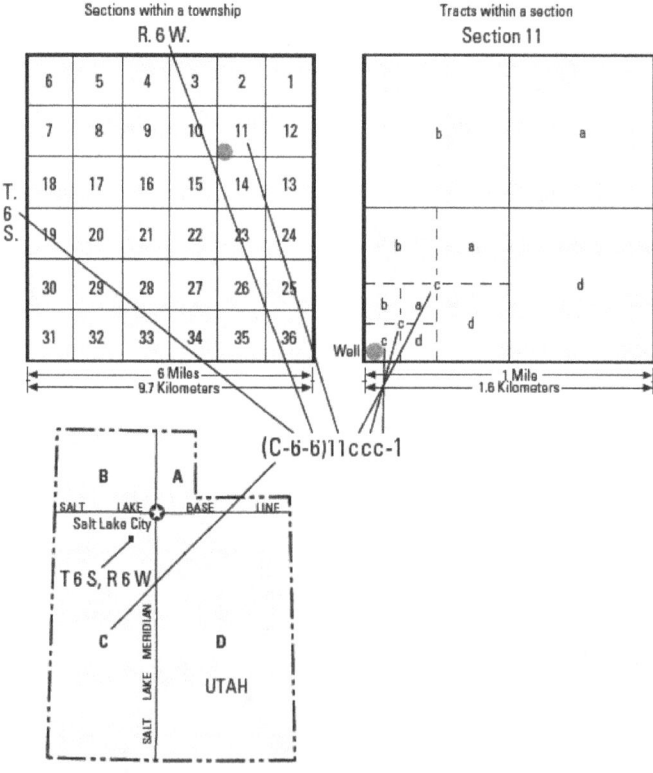

[1]Although the basic land unit, the section, is theoretically 1 square mile, many sections are irregular. Such sections are subdivided into 10-acre tracts, generally beginning at the southeast corner, and the surplus or shortage is taken up in the tracts along the north and west sides of the section.

Hydrogeologic and Geochemical Characterization of Groundwater Resources in Rush Valley, Tooele County, Utah

By Philip M. Gardner and Stefan Kirby

Abstract

The water resources of Rush Valley were assessed during 2008–2010 with an emphasis on refining the understanding of the groundwater-flow system and updating the groundwater budget. Surface-water resources within Rush Valley are limited and are generally used for agriculture. Groundwater is the principal water source for most other uses including supplementing irrigation. Most groundwater withdrawal in Rush Valley is from the unconsolidated basin-fill aquifer where conditions are generally unconfined near the mountain front and confined at lower altitudes near the valley center. Productive aquifers also occur in fractured bedrock along the valley margins and beneath the basin-fill deposits in some areas.

Drillers' logs and geophysical gravity data were compiled and used to delineate seven hydrogeologic units important to basin-wide groundwater movement. The principal basin-fill aquifer includes the unconsolidated Quaternary-age alluvial and lacustrine deposits of (1) the upper basin-fill aquifer unit (UBFAU) and the consolidated and semiconsolidated Tertiary-age lacustrine and alluvial deposits of (2) the lower basin-fill aquifer unit (LBFAU). Bedrock hydrogeologic units include (3) the Tertiary-age volcanic unit (VU), (4) the Pennsylvanian- to Permian-age upper carbonate aquifer unit (UCAU), (5) the upper Mississippian- to lower Pennsylvanian-age upper siliciclastic confining unit (USCU), (6) the Middle Cambrian- to Mississippian-age lower carbonate aquifer unit (LCAU), and (7) the Precambrian- to Lower Cambrian-age noncarbonate confining unit (NCCU). Most productive bedrock wells in the Rush Valley groundwater basin are in the UCAU.

Average annual recharge to the Rush Valley groundwater basin is estimated to be about 39,000 acre-feet. Nearly all recharge occurs as direct infiltration of snowmelt and rainfall within the mountains with smaller amounts occurring as infiltration of streamflow and unconsumed irrigation water at or near the mountain front. Groundwater generally flows from the higher altitude recharge areas toward two distinct valley-bottom discharge areas: one in the vicinity of Rush Lake in northern Rush Valley and the other located west and north of Vernon. Average annual discharge from the Rush Valley groundwater basin is estimated to be about 43,000 acre-feet. Most discharge occurs as evapotranspiration in the valley lowlands, as discharge to springs and streams, and as withdrawal from wells. Subsurface discharge outflow to Tooele and Cedar Valleys makes up only a small fraction of natural discharge.

Groundwater samples were collected from 25 sites (24 wells and one spring) for geochemical analysis. Dissolved-solids concentrations in water from these sites ranged from 181 to 1,590 milligrams per liter. Samples from seven wells contained arsenic concentrations that exceed the Environmental Protection Agency Maximum Contaminant Level of 10 micrograms per liter. The highest arsenic levels are found north of Vernon and in southeastern Rush Valley. Stable-isotope ratios of oxygen and deuterium, along with dissolved-gas recharge temperatures, indicate that nearly all modern groundwater is meteoric and derived from the infiltration of high altitude precipitation in the mountains. These data are consistent with recharge estimates made using a Basin Characterization Model of net infiltration that shows nearly all recharge occurring as infiltration of precipitation and snowmelt within the mountains surrounding Rush Valley. Tritium concentrations between 0.4 and 10 tritium units indicate the presence of modern (< 60 years old) groundwater at 7 of the 25 sample sites. Apparent $^3H/^3He$ ages, calculated for six of these sites, range from 3 to 35 years. Adjusted minimum radiocarbon ages of premodern water samples range from about 1,600 to 42,000 years with samples from 11 of 13 sites being more than 11,000 years. These data help to identify areas where modern groundwater is circulating through the hydrologic system on time scales of decades or less and indicate that large parts of the principal basin-fill and the bedrock aquifers are much less active and receive little to no modern recharge.

Introduction

Rush Valley is a rural valley located in west-central Utah about 50 mi west-southwest of Salt Lake City (fig. 1). It is a typical Basin and Range valley bounded by mountains with large range-front faults along its margins and is characterized by varying thickness of basin fill within the down-dropped valley basin. Surface-water resources are limited because few perennial streams enter the valley from the mountains and no streams flow through or exit the valley. Although the limited surface-water resources generally are used for agriculture (and irrigation is supplemented with groundwater), groundwater is the predominant source of water for most other uses. Hood and others (1969) estimated annual groundwater recharge and discharge in Rush Valley to be about 34,000 and 37,000 acre-feet (acre-ft), respectively.

Tooele Valley to the north and Cedar Valley to the east of Rush Valley (fig. 1) have experienced significant suburban growth in step with the westward expansion of the Salt Lake urban corridor. Both of these valleys have limited water resources and are closed to new groundwater appropriations. Previous studies have shown the three valleys to be hydraulically connected, with groundwater movement in the basin-fill aquifer out of Rush Valley and into Tooele and Cedar Valleys, both at lower altitudes (Hood and others, 1969; Razem and Steiger, 1981; Stolp, 1994; Lambert and Stolp, 1999; Stolp and Brooks, 2009). Development of groundwater resources in Rush Valley recently has been proposed to supply water to the growing populations in Tooele and Cedar Valleys.

Prior to the late 1960s, few published sources of hydrologic data in Rush Valley were available and most of those sources discussed the valley only as part of a broader area (Carpenter, 1913; Mahoney, 1953; Snyder, 1963; Bagley and others, 1964). Gates (1963, 1965) compiled basic data and described groundwater conditions along the boundary between Rush and Tooele Valleys. Feltis (1967) discussed recharge conditions in the Oquirrh Mountains in a report describing groundwater conditions in Cedar Valley. A hydrologic reconnaissance of Rush Valley was conducted in 1967–1968 (Hood and others, 1969). This study developed a general water budget for the valley and compiled available water-level, water-quality, groundwater-withdrawal, aquifer-property, and surface-water data. Since then, additional hydrologic data have been collected as part of various monitoring programs or site-specific studies. In November 1994, a 9-day multiple-well aquifer test was conducted by the U.S. Geological Survey (USGS) in the Vernon area, and the unpublished results of that test are included in the Aquifer Properties section of this report. A separate study by the USGS focused on monitoring groundwater conditions in the area around Clover Creek from 1995 through 1999 (M. Enright, written commun., 1997, 1999). A long-term monitoring and sampling program began in 1998 to evaluate groundwater movement and water quality at the Deseret Chemical Depot (DCD), which occupies approximately 30 mi^2 in east-central Rush Valley (North Wind Inc., 2008). Water-level and groundwater-geochemistry data have been collected from northern Rush Valley as part of several USGS studies of Tooele Valley (Stolp, 1994; Lambert and Stolp, 1999; Stolp and Brooks, 2009). Jordan and Sabah of the Utah Geological Survey are currently investigating the hydrology of Cedar Valley to the east of Rush Valley and developing a numerical groundwater-flow model for that valley that incorporates an estimate of subsurface flow from Rush Valley to Cedar Valley (L. Jordan, oral commun., 2010).

Purpose and Scope

The purpose of this report is to describe the hydrology and groundwater resources of Rush Valley and to present a revised conceptual model of the groundwater-flow system. Emphasis is placed on refining the hydrologic concepts presented by Hood and others (1969). Specifically, the current study presents updated groundwater budgets, a new regional map of the groundwater-level surface, a new overview of groundwater geochemistry/water quality in the basin-fill and carbonate aquifers, a revised hydrogeologic framework of the basin-fill and consolidated-rock aquifers, and a reassessment of interbasin flow from Rush Valley to adjacent valleys. Water-level fluctuations and a characterization of groundwater-flow paths also are described and evaluated. Surface water is described primarily as it relates to groundwater recharge and discharge. This study includes a description of groundwater recharge using the Basin Characterization Model (BCM; Flint and Flint, 2007), a recently developed distributed net-infiltration approach. Major components of groundwater discharge were reestimated using up-to-date information on land-use and water consumption by natural vegetation. Environmental tracers were used to assess groundwater-recharge sources, flow directions, and residence times. Results of the study are intended to improve the understanding of the Rush Valley groundwater-flow system and the ability of water managers to assess the long-term sustainability of current and future groundwater supplies developed in Rush Valley and adjacent valleys.

Physical Characteristics of the Study Area

Rush Valley is located in the Basin and Range physiologic province (Fenneman, 1931) and exhibits geologic and topographic characteristics typical of the region. The study area is defined by the surrounding topographic divide corresponding to the eight-digit Hydrologic Unit Code (HUC) 16020304 (Watershed Boundary Dataset, 2007). It encompasses approximately 720 mi^2 and is located in a part of the Great Basin that once was occupied by Lake Bonneville (Gilbert, 1890). The basin-fill part of Rush Valley occupies an area of about 510 mi^2 at altitudes between about 5,000 and 6,000 ft and is surrounded by the contributing mountain watersheds that make up the remaining 210 mi^2. Sediment eroded from the surrounding mountains is the source of the Tertiary- and Quaternary-age deposits that comprise the

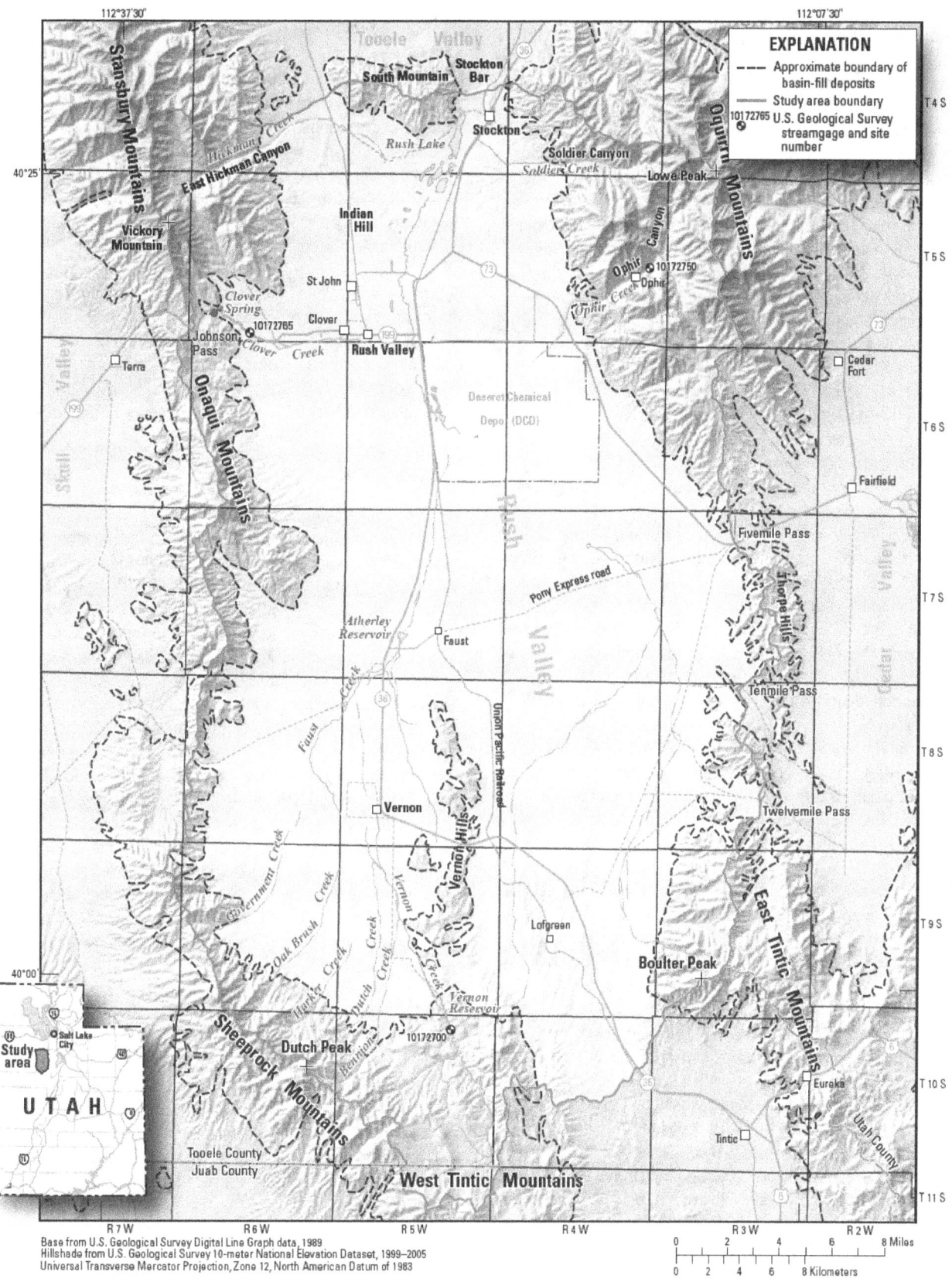

Figure 1.   Location of the Rush Valley study area, Tooele County, Utah.

basin, forming the valley floor. Rush Valley is bounded by South Mountain to the north; the Sheeprock and West Tintic Mountains to the south; the Oquirrh Mountains, Thorpe Hills, and East Tintic Mountains to the east; and the Stansbury and Onaqui Mountains to the west (fig. 1). The mountains that surround Rush Valley are composed of folded and faulted blocks of Precambrian- through Paleozoic-age sedimentary rocks and localized igneous rocks (Hood and others, 1969).

Altitudes in Rush Valley range from about 5,000 ft at Rush Lake to over 10,000 ft in the highest parts of the Oquirrh and Stansbury Mountains. The highest point in the study area is Lowe Peak (10,572 ft) in the Oquirrh Mountains (fig. 1). Vickory Mountain (10,305 ft) in the southern Stansbury Mountains and Dutch Peak (8,964 ft) in the Sheeprock Mountains are other notable high points along the divide. The lowest point (5,175 ft) along the study area boundary is a large and well-preserved shoreline remnant of Lake Bonneville named the Stockton Bar, located just north of the town of Stockton and east of South Mountain. Rush Valley is higher than the valleys that surround it on all sides. Consequently, it was only partially inundated by a shallow, sheltered arm of Lake Bonneville.

Hood and others (1969) describe three morphologically distinct areas of Rush Valley. Northern Rush Valley (north of about T. 6 S.; fig. 1) is characterized by broad alluvial fans that descend abruptly from the mountain front and then more gently to the valley floor. The alluvial fans are deeply incised by stream channels from large canyons draining the Stansbury and Oquirrh Mountains. Drainage is generally toward the flat and smooth central valley floor intermittently occupied by Rush Lake.

Southern Rush Valley can be divided into two parts that are separated by the Vernon Hills, a low topographic divide that extends northward from the West Tintic Mountains. West of this divide is the Vernon area, a subbasin occupied by the town of Vernon and surrounding agricultural lands. The mountains bordering this area are not as high as those in northern Rush Valley and receive less precipitation. As a result, the alluvial slopes coming off of the Sheeprock and West Tintic Mountains are not as deeply incised as those in the north. Drainage is toward Vernon and Government Creeks in the center of this subbasin and then northward toward Faust Creek.

Southeastern Rush Valley is the driest part of the study area. The mountains surrounding it are lower in altitude and receive less precipitation than the mountains bordering other parts of Rush Valley. Much of this area is composed of unconsolidated deposits that thinly mantle older semiconsolidated basin-fill deposits. Drainage is northeastward toward a string of connected playas centered in T. 7 S., R. 4 W (Hood and others, 1969). Unlike the lowland parts of northern Rush Valley and the Vernon area, the water table is well below land surface and the playas are not active areas of groundwater discharge.

Population and Land Use

The population of Rush Valley is approximately 1,100 (U.S. Census Bureau, 2000) and has not changed substantially since 1960 (Hood and others, 1969). Some additional residential development has occurred since 2000 in the northern part of the valley along the south slope of South Mountain. Communities in the valley include Stockton, the town of Rush Valley (previously Clover / St. John), Vernon, the DCD, and the small mountain town of Ophir (fig. 1).

Land use within the Rush Valley study area includes the industrial and military operations of the DCD; irrigated and nonirrigated farmland and pasture used for agriculture and livestock grazing; incorporated and unincorporated residential areas; and mining. The mountain areas remain mostly undeveloped and are used primarily for recreation, mining, and grazing. The central lowlands of northern Rush Valley and the area around Vernon contain sizable areas of wetlands.

Precipitation

The average annual precipitation (1971–2000) estimated from Parameter-elevation Regressions on Independent Slopes Model (PRISM) data (Daly and others, 2008) in the Rush Valley drainage basin ranges from less than 12 in. in the central lowlands of the valley bottom to more than 40 in. at the highest altitudes in the Stansbury and Oquirrh Mountains (fig. 2). Most precipitation occurs during the winter and early spring months as snowfall and the least occurs during July and August. Three weather stations located in or near Rush Valley with long-term records of precipitation illustrate the variation in annual rainfall (fig. 3). All three stations recorded multiple extended periods (greater than about 5 years) of below average precipitation (1972–1976 and 1999–2003) and above average precipitation (1980–1986 and 1993–1998) since about 1970. Only the Fairfield weather station has a precipitation record dating back long enough to capture the southwestern regional drought lasting from about 1953–1965 (U.S. Geological Survey, 1991).

Although annual precipitation rates are much higher in the mountains surrounding Rush Valley, the low-altitude valley area is so large that it receives the majority of annual rainfall, about 394,000 acre-ft out of about 661,000 acre-ft (1971–2000 average estimated from PRISM model data (Daly and others, 2008). It is estimated that all of the precipitation that falls at valley altitudes (below about 5,500–6,000 ft) is consumed by evapotranspiration (ET). Precipitation falling at higher altitudes generally exceeds the amount consumed by ET and becomes either direct infiltration or runoff in streams draining the mountains.

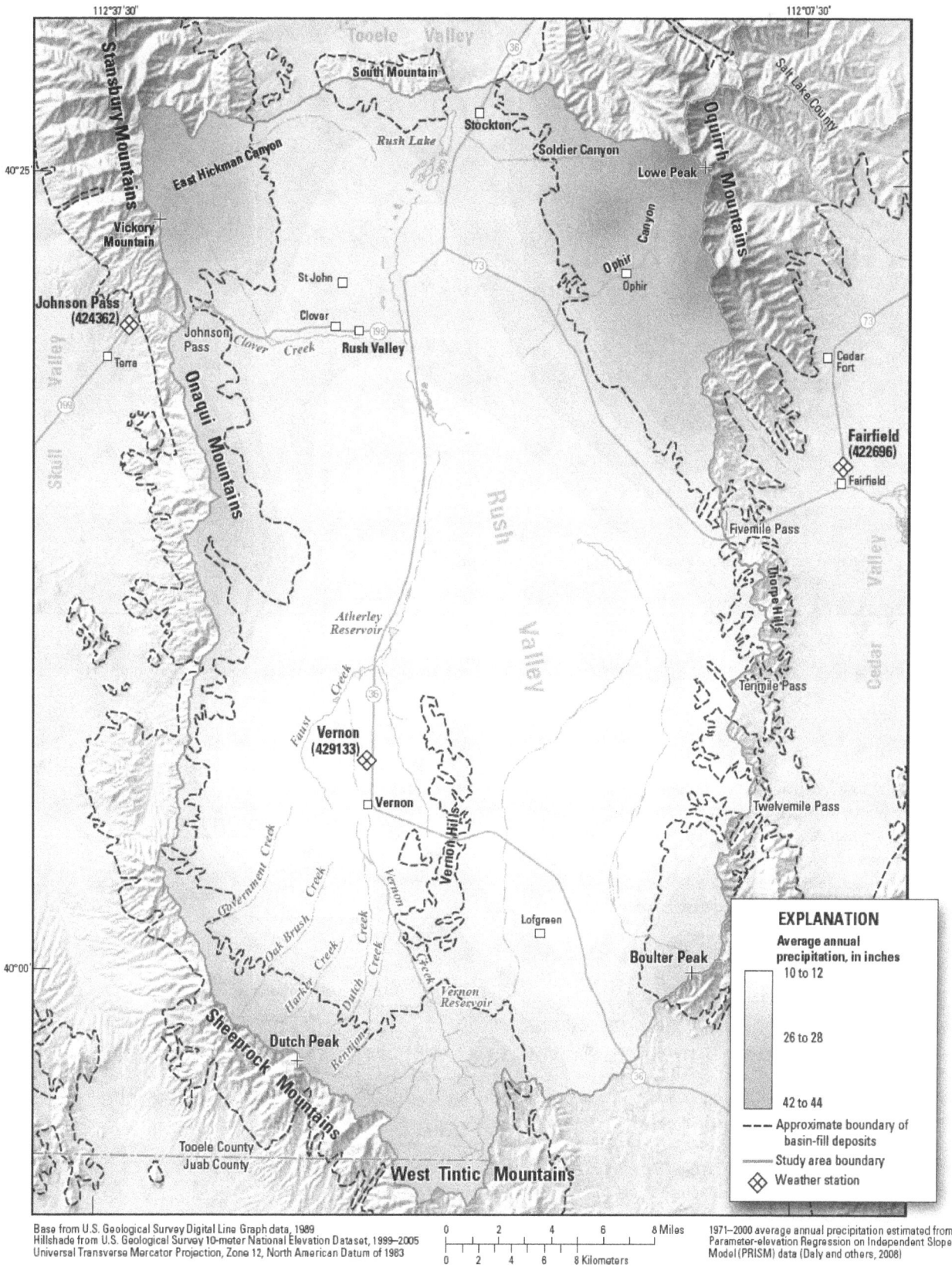

Figure 2.   Average annual precipitation, 1971–2000, Rush Valley, Tooele County, Utah.

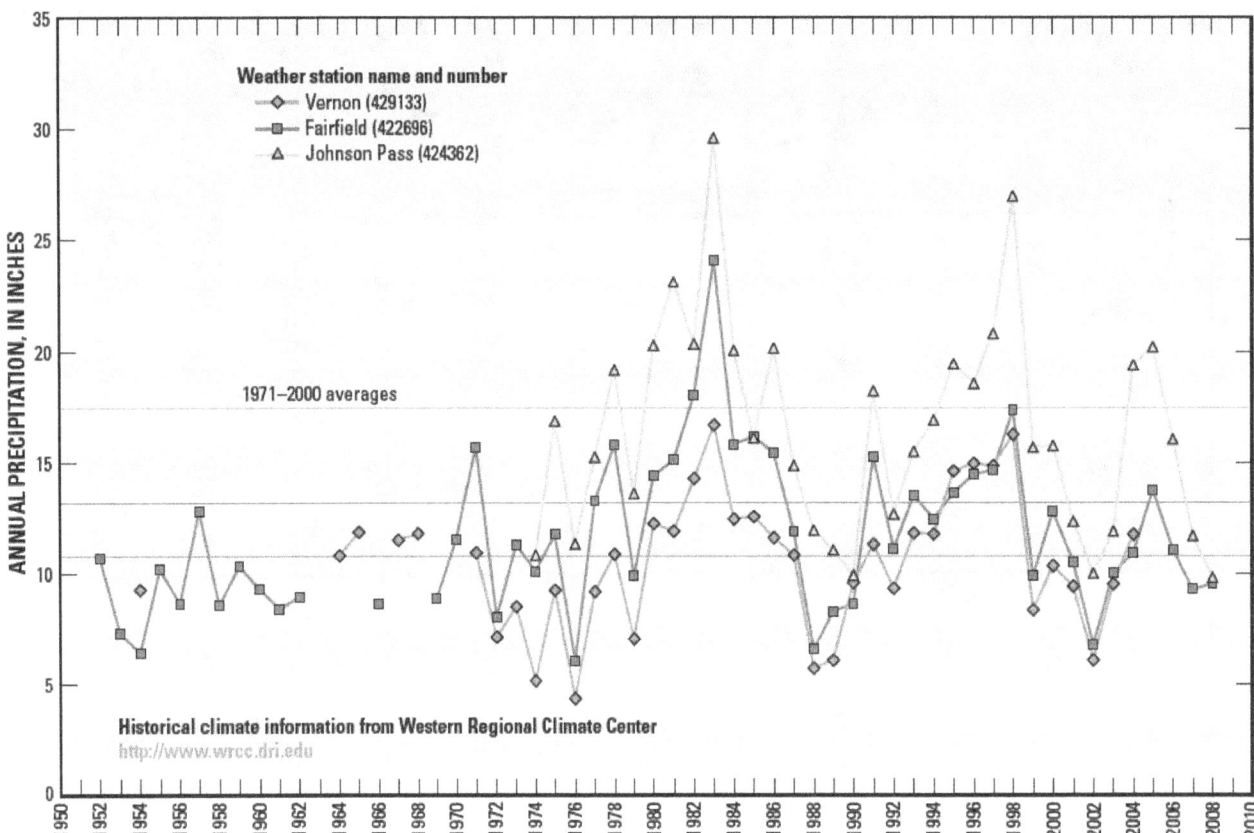

Figure 3.   Long-term annual precipitation recorded at three weather stations in or near Rush Valley, Tooele County, Utah.

Streamflow

Most streams in Rush Valley are intermittent and flow only in response to periods of snowmelt or intense rainfall. Much of the water in these intermittent streams is lost to evapotranspiration or infiltration on the alluvial slopes of the valley. Small amounts of streamflow occasionally reach the valley bottom at playas in the east-central part of the valley or at Rush Lake where it is evaporated.

Nine perennial streams drain the mountains surrounding Rush Valley, and annual streamflow ranges from about 400 to 7,000 acre-ft (table 1). High flow in these streams occurs during the spring months when they collect runoff from the melting high-altitude snowpack. The headwaters of Ophir and Soldier Creeks are located in the southern Oquirrh Mountains on the east side of northern Rush Valley (fig. 1). The headwaters of Clover and Hickman Creeks are located in the southern Stansbury Mountains on the west side of northern Rush Valley. The headwaters of Vernon, Bennion, Dutch, Harker, and Oak Brush Creeks are located in the Sheeprock Mountains south of Vernon. All of the perennial streams are sustained year round by groundwater discharge either from mountain springs or as baseflow directly to the stream. Water from each of the named streams is captured upstream of the canyon mouths and delivered for irrigation or domestic use

Table 1.   Average annual streamflow, 1971–2000, in nine perennial streams that drain the mountains surrounding Rush Valley, Tooele County, Utah.

[All flows in acre-feet per year rounded to the nearest 100]

Stream name	Mountain range	Average annual streamflow	Fraction of streamflow that is base flow	Base flow
[1]Ophir Creek	Oquirrh	[1]7,000	0.33	2,300
[1]Soldier Creek	Oquirrh	[1]2,400	0.33	790
[1]Hickman Creek	Stansbury	[1]4,800	0.48	2,300
[2]Clover Creek	Stansbury	[2]3,400	0.48	1,600
[3]Vernon Creek	Sheeprock	[3]3,200	0.79	2,500
[4]Harker Creek	Sheeprock	[4]700	0.79	550
[4]Bennion Creek	Sheeprock	[4]600	0.79	470
[4]Dutch Creek	Sheeprock	[4]400	0.79	320
[4]Oak Brush Creek	Sheeprock	[4]400	0.79	320
Total		22,900		11,150

[1]Regression-estimated value

[3]Based on 1986–2001 water-year record at gage 10172765

[3]Based on 1971–2000 water-year record at gage 10172700

[4]55 percent of the regression-estimated value, as described in report

by water or irrigation companies with surface-water rights in Rush Valley. Except during periods of extremely high runoff, when streamflow exceeds what is captured, no water from the perennial mountain streams reaches the valley floor.

Three of the nine perennial streams have some period of recorded streamflow. Streamflow in Ophir Creek (USGS Gage 10172750) was measured continuously from March 1986 through April 1987. Clover Creek (USGS Gage 10172765) was measured continuously from December 1984 to October 2001. Vernon Creek (USGS Gage 10172700) has been monitored continuously since 1958.

Average annual (1971–2000) streamflow was estimated for seven of the nine perennial streams having insufficient or no streamflow record (table 1) using regional regression equations that were developed to predict mean annual streamflow at ungaged sites in Utah (Wilkowske and others, 2008). Because the regression equations used to calculate annual streamflow are based only on basin size and average annual precipitation, and because geologic characteristics in the mountains appear to influence patterns of runoff versus infiltration, regression estimates were compared to gage data in drainages where they were available. The average streamflow for the 1 year of measurement in Ophir Creek (1986–87) was 6,700 acre-ft. This measurement occurred during a period of approximately normal precipitation (fig. 3) and compared well with the regression-estimated average streamflow of 7,000 acre-ft. The regression-estimated annual streamflow for Soldier Creek is 2,400 acre-ft and is assumed to be reasonable based on having similar drainage characteristics as Ophir Creek, also in the Oquirrh Mountains.

The average annual streamflow in Clover Creek is 3,400 acre-ft, based on 1986–2001 gaged measurements. This is substantially more than the regression-estimated value of 2,200 acre-ft (fig. 4). It is believed that the geologic structure of the southern Stansbury Mountains is responsible for the headwater spring (Clover Spring) of this stream having a higher discharge than expected given the size of its surface-water drainage basin. Steeply dipping bedding and a high-angle thrust fault, located northeast of Clover Spring, appear to act as a barriers to mountain groundwater that would otherwise flow east, forcing more groundwater in the southern end of the Stansbury Mountains to flow south and discharge at Clover Spring or as baseflow in Clover Creek. Hickman Creek, also emanating from the Stansbury Mountains, has never been gaged; its regression-estimated average annual streamflow is 4,800 acre-ft.

The gage at Vernon Creek, located in the Sheeprock Mountains south of Vernon, has more than a 50-year record of streamflow. The average annual (1958–2009) streamflow of Vernon Creek, located in the Sheeprock Mountains south of Vernon is 2,700 acre-ft. However, the 1971–2000 average annual streamflow was compared to the regression-estimated streamflow because it is based on 1971–2000 PRISM average precipitation. The average annual streamflow of Vernon Creek for 1971–2000 is 3,200 acre-ft, which is only 55 percent of the regression-estimated streamflow of 5,800 acre-ft (fig. 4).

This discrepancy may be due to the localized permeable nature of exposed bedrock in the Sheeprock Mountains, permitting a higher fraction of precipitation to infiltrate and become groundwater recharge than that estimated by regression. Assuming that these characteristics are similar throughout the Sheeprock Mountains, average annual streamflow for Bennion, Dutch, Harper, and Oak Creeks was assumed to equal 0.55 times the regression-estimated value (table 1).

The amount of baseflow versus overland runoff that makes up annual streamflow in a drainage varies depending mostly on the geology of the watershed (Tague and Grant, 2004; Gardner and others, 2010). High-permeability materials in the recharge area result in large volumes of snowmelt infiltration, sustaining the discharge of springs and stream baseflow throughout the year. Streams in watersheds dominated by less permeable geologic formations have higher amplitude runoff peaks due to increased over-land runoff fractions. Baseflow was estimated for streams with gage data by assuming that the average daily discharge for the low-flow months of December, January, and February represented constant baseflow throughout the year. The volume of estimated baseflow was compared to the total annual streamflow and this fraction was assumed to be the same for streams in nearby drainages with similar geology (table 1). This is a conservative approach that may underestimate annual baseflow as it is likely that the baseflow fraction of total discharge increases by an unknown amount during periods of high flow.

Geology

Rush Valley is a large, north-south trending, internally drained basin that is defined by a series of narrow, normal-fault-bounded bedrock mountain ranges and adjoining low hills that surround a broad, gently sloping valley floor. Bedrock in the mountains and hills surrounding and within Rush Valley is characterized by a thick section of complexly faulted and folded Precambrian through Paleozoic-age sedimentary rocks that include carbonate rocks (limestone and dolomite), quartzite, sandstone, and shale (Hintze and others, 2000; Clark and others, 2009). Basin fill has been deposited between prominent mountain ranges and covers most of the floor of Rush Valley. Basin fill in Rush Valley includes a range of semiconsolidated to unconsolidated lacustrine and alluvial or colluvial deposits and lesser amounts of extrusive volcanic rocks (Hood and others, 1969; Everitt and Kaliser, 1980). Unconsolidated basin fill, and to a lesser degree, Pennsylvanian-age Oquirrh Group bedrock comprise the principal aquifers in Rush Valley (Hood and others, 1969).

The early tectonic history of Rush Valley is recorded by exposed Precambrian to Pennsylvanian-age strata that were deposited first across a broad subsiding marine platform and later within the rapidly subsiding Oquirrh Basin. Rocks deposited during this period include 1) Precambrian to Early Cambrian-age quartzite, shale, and conglomerate, 2) a thick sequence of Middle Cambrian to Mississippian-age limestone, dolomite, sandstone, shale, and quartzite, and 3)

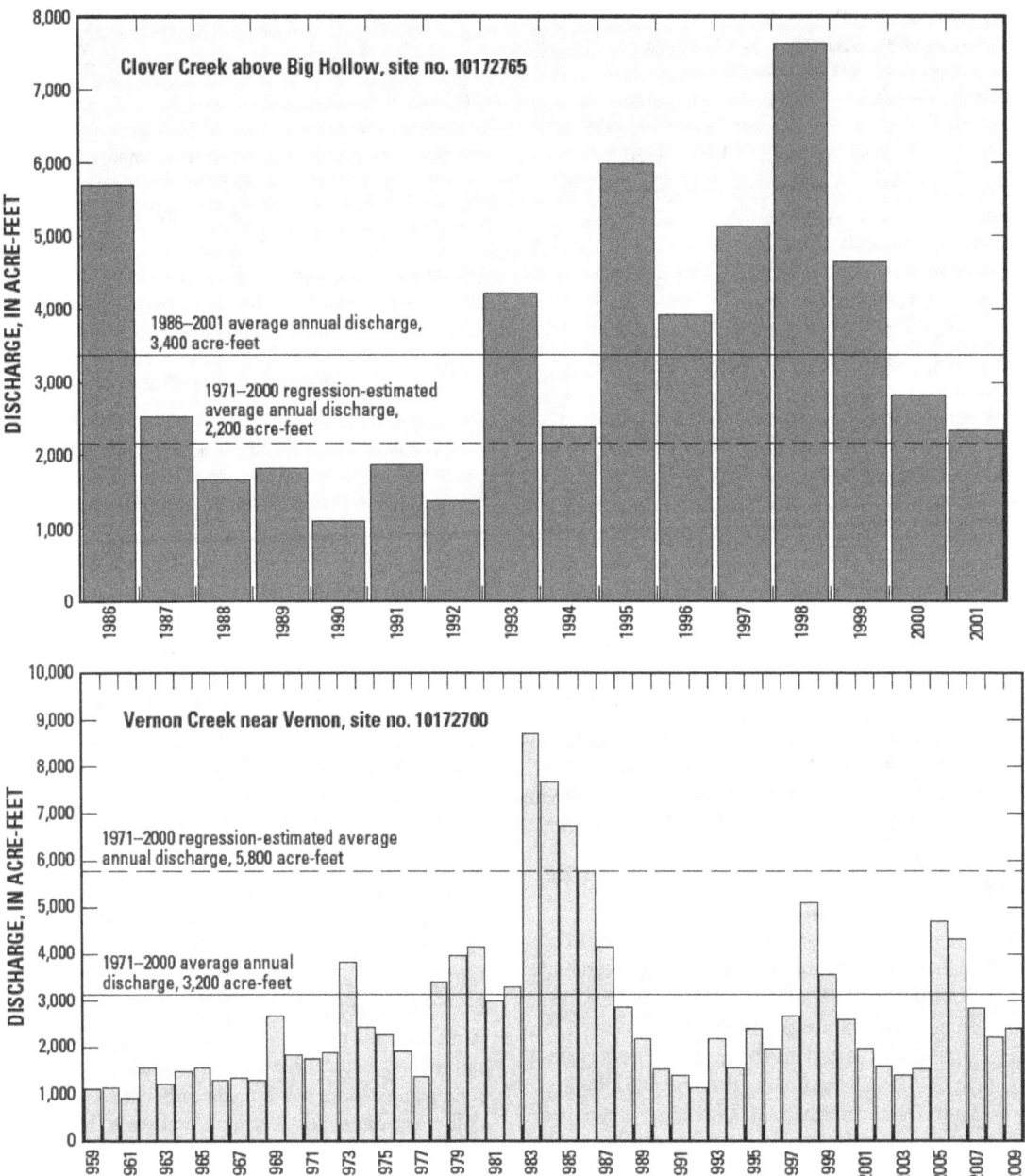

Figure 4.   Annual streamflow of Clover Creek for 1986–2001 and Vernon Creek for 1959–2009, Rush Valley, Tooele County, Utah.

Pennsylvanian to Early Permian-age interbedded limestone, sandstone, and quartzite. These rocks were folded and faulted by dominantly east-directed thrust faulting and compression during the Late Jurassic to Eocene-age Sevier orogenic event (Armstrong, 1968; DeCelles and Coogan, 2006).

During the Eocene Epoch, crustal shortening was replaced by roughly east-west extension and significant regional volcanism (Constenius, 1996; Constenius and others, 2003). Early extension, localized along north-south-striking normal faults, controlled the formation of narrow, rapidly subsiding basins into which sediment from surrounding uplands and nearby volcanic centers was deposited. Basin-fill includes 1)

Eocene or Oligocene-age extrusive tuffaceous volcanics, 2) consolidated to semiconsolidated Miocene-age tuffaceous lacustrine and alluvial deposits, and 3) unconsolidated latest Tertiary to Quaternary-age alluvial, colluvial, and lacustrine deposits (Everitt and Kaliser, 1980). Extension remains the dominant tectonic force in the area, but has varied in magnitude, style, and extent during Eocene to Holocene time. Fault scarps that displace Quaternary-age unconsolidated deposits parallel mountain fronts in Rush Valley (Everitt and Kaliser, 1980; Clark and others, 2009) and provide examples of Holocene-age extension in this part of the Basin and Range.

Groundwater Hydrology

Groundwater is the primary source of drinking water in northern Rush Valley and also is used for irrigation, stock watering, and industrial purposes. Productive aquifers are present in both bedrock and unconsolidated basin-fill deposits. The majority of wells in the study area are completed within the basin-fill deposits because of the ease of drilling, accessibility, and proximity to populated areas. While the basin-fill aquifer has been the primary source of groundwater for more than a half century, bedrock wells within the study area are increasingly being developed, especially beneath the basin fill near Vernon, along the south flank of South Mountain, and in the Oquirrh Mountains.

Large parts of the Rush Valley groundwater basin are conceptualized as a single, interconnected hydrologic system where recharge that occurs through consolidated rock in the mountains can enter and move through the adjoining basin-fill deposits in the valleys. In this conceptualization, groundwater withdrawn from bedrock or basin-fill aquifers downgradient of recharge areas is replenished at the average annual recharge rate. There are, however, areas where productive aquifers appear to be disconnected from the active groundwater system and are isolated from modern recharge.

Hydrogeology

The geologic units in the study area both store and convey groundwater and impart control on regional groundwater movement. The geologic complexity, scale of the study area, and potential for groundwater to reside in and travel through multiple geologic units necessitate generalization of the aquifer system. For this study, geologic units are grouped into seven hydrogeologic units on the basis of their relevance to basin-wide groundwater movement (fig. 5). The names given to the hydrogeologic units in Rush Valley are the same as those used in a recent study of the Great Basin carbonate and alluvial aquifer system (Sweetkind and others, 2011), covering an area that encompasses Rush Valley and has undergone a common geologic evolution. Distinct basin-fill units include (1) an upper basin-fill aquifer unit (UBFAU), heterogeneous but laterally extensive unconsolidated Quaternary-age alluvial and lacustrine, deposits and (2) a lower basin-fill aquifer unit (LBFAU), a thick sequence of consolidated and semiconsolidated Tertiary-age lacustrine and alluvial deposits of the Salt Lake Formation (fig. 5). Regionally important bedrock units include (1) an upper carbonate aquifer unit (UCAU), interbedded Pennsylvanian to Permian-age carbonates, sandstone, and quartzite collectively referred to as the Oquirrh Group, (2) an upper siliciclastic confining unit (USCU), upper Mississippian to lower Pennsylvanian-age shaly, fine-grained sedimentary rocks comprised mostly of the Manning Canyon Shale, (3) a lower carbonate aquifer unit (LCAU), sequences of middle Cambrian to Mississippian-age carbonates, sandstone, quartzite, and shale, (4) a noncarbonate

confining unit (NCCU), a thick section of Precambrian to Lower Cambrian-age sedimentary rocks comprised mostly of quartzite, shale, and diamictite, and (5) a volcanic unit (VU), undifferentiated Tertiary-age volcanic rocks that are of limited extent within the study area.

The hydrogeologic units in the study area form two distinct aquifer systems composed of alternating more permeable and less permeable units. The two general types of aquifer materials are: permeable parts of the upper and lower basin-fill aquifer units (UBFAU and LBFAU) and carbonate rocks of the upper and lower carbonate aquifer units (UCAU and LCAU). Each of these units may include one or more water-bearing zones but are stratigraphically and structurally heterogeneous, resulting in a highly variable ability to store and transmit water.

Basin Fill

The primary basin-fill aquifer (subsequently referred to as "basin-fill aquifer" or "basin fill") in Rush Valley is a lithologically diverse group of sediments deposited during normal faulting and formation of the basin and range topography that has occurred since the middle Tertiary. The basin fill is subdivided into two hydrogeologic units: the UBFAU, young unconsolidated deposits, and the LBFAU, older basin fill that includes the Salt Lake Formation and other consolidated and semiconsolidated deposits.

The unconsolidated deposits of the UBFAU consist of gravel, sand, silt, and clay deposited along alluvial fans and channels that extend from the mountain ranges and cover most of the floor of Rush Valley (fig. 5). Moving laterally from the valley margin toward the center of the valley, coarse-grained material grades into or overlies fine-grained sand, silt, clay, and marl deposited during episodic lacustrine conditions. On the basis of well logs, total thickness of the UBFAU may exceed 1,000 ft in places (fig.6B, section D–D'). More commonly these deposits are 300 to 500 ft thick (figs. 6A and B). In the southeastern part of Rush Valley, the UBFAU is less than 300 ft thick and overlies the older Salt Lake Formation of the LBFAU (fig.6B, section E–E'). The UBFAU contains spatially extensive clay layers in several areas including near Vernon and near Clover and Saint John (table 2). Where present, these clay layers may form important confining layers within the basin-fill aquifer. Elsewhere, especially near the valley margins, unconfined or partially confined conditions are typical in much of the UBFAU. Permeability of the UBFAU is highly variable but generally decreases toward fine-grained deposits along the central axis of Rush Valley.

Well logs and gravity data imply the presence of older, consolidated to semiconsolidated LBFAU deposits beneath large parts of Rush Valley. The oldest basin fill in Rush Valley consists of tuffaceous lacustrine and alluvial deposits that comprise the LBFAU. Outcrops of this unit are found near the north end of Vernon Hills and along the Pony Express Road (fig. 5); elsewhere the unit is overlain by the younger basin-fill deposits of the UBFAU. The LBFAU may be up to 4,500 ft

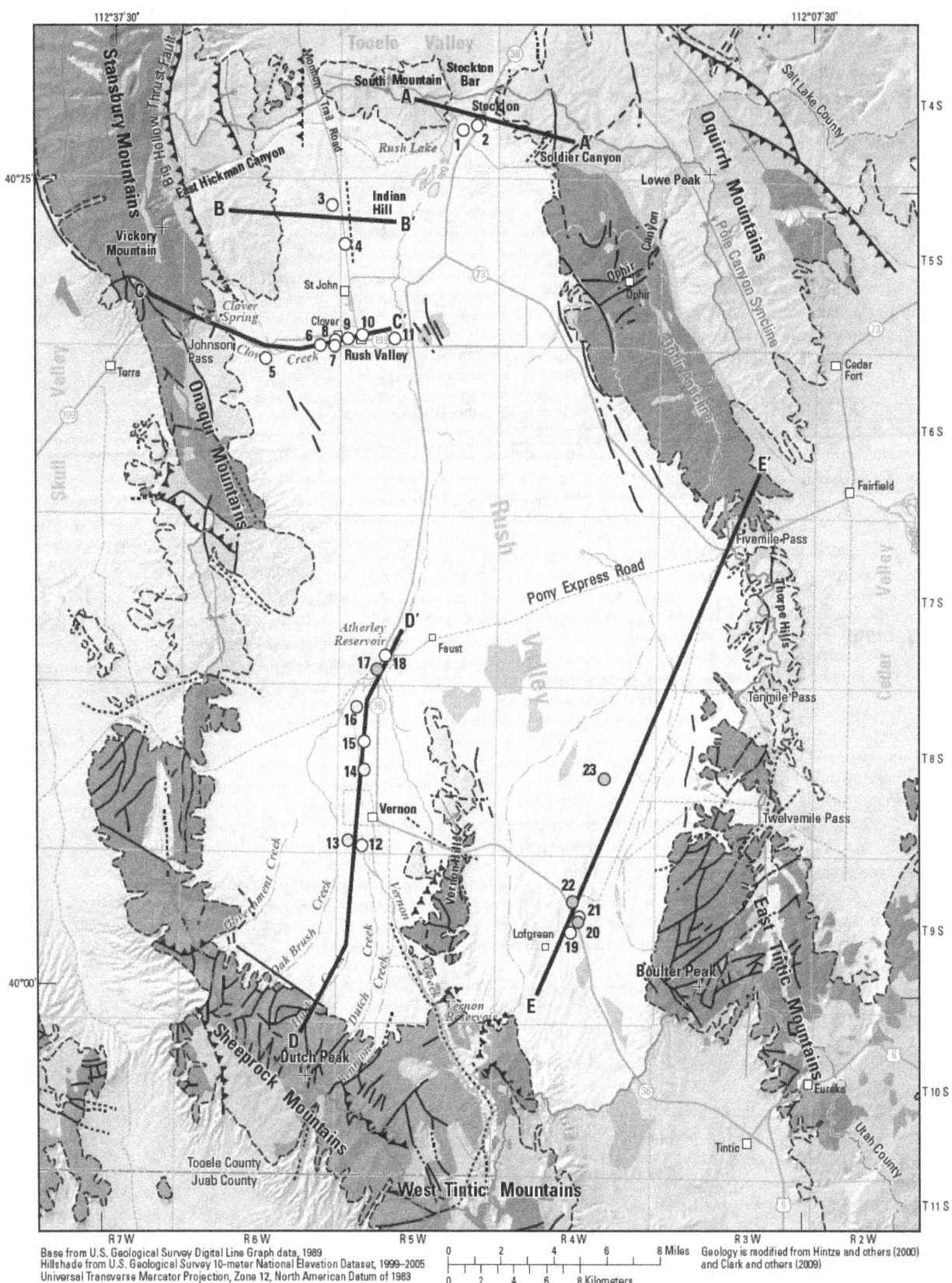

Figure 5. Surficial extent of hydrogeologic units and prominent structural geologic features of the Rush Valley area, Tooele County, Utah.

EXPLANATION

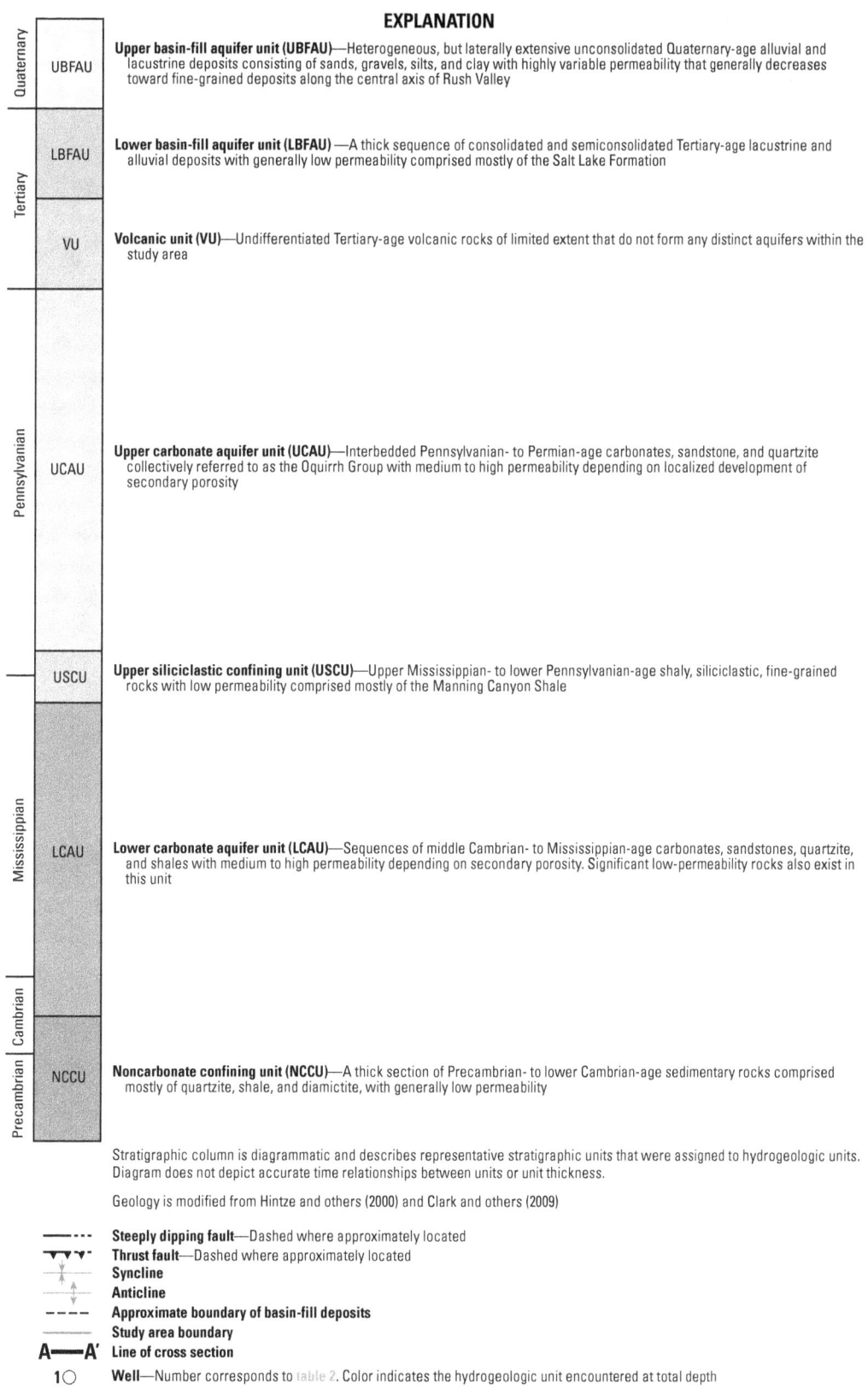

Upper basin-fill aquifer unit (UBFAU)—Heterogeneous, but laterally extensive unconsolidated Quaternary-age alluvial and lacustrine deposits consisting of sands, gravels, silts, and clay with highly variable permeability that generally decreases toward fine-grained deposits along the central axis of Rush Valley

Lower basin-fill aquifer unit (LBFAU)—A thick sequence of consolidated and semiconsolidated Tertiary-age lacustrine and alluvial deposits with generally low permeability comprised mostly of the Salt Lake Formation

Volcanic unit (VU)—Undifferentiated Tertiary-age volcanic rocks of limited extent that do not form any distinct aquifers within the study area

Upper carbonate aquifer unit (UCAU)—Interbedded Pennsylvanian- to Permian-age carbonates, sandstone, and quartzite collectively referred to as the Oquirrh Group with medium to high permeability depending on localized development of secondary porosity

Upper siliciclastic confining unit (USCU)—Upper Mississippian- to lower Pennsylvanian-age shaly, siliciclastic, fine-grained rocks with low permeability comprised mostly of the Manning Canyon Shale

Lower carbonate aquifer unit (LCAU)—Sequences of middle Cambrian- to Mississippian-age carbonates, sandstones, quartzite, and shales with medium to high permeability depending on secondary porosity. Significant low-permeability rocks also exist in this unit

Noncarbonate confining unit (NCCU)—A thick section of Precambrian- to lower Cambrian-age sedimentary rocks comprised mostly of quartzite, shale, and diamictite, with generally low permeability

Stratigraphic column is diagrammatic and describes representative stratigraphic units that were assigned to hydrogeologic units. Diagram does not depict accurate time relationships between units or unit thickness.

Geology is modified from Hintze and others (2000) and Clark and others (2009)

- - - **Steeply dipping fault**—Dashed where approximately located
▼▼ **Thrust fault**—Dashed where approximately located
Syncline
Anticline
- - - - **Approximate boundary of basin-fill deposits**
Study area boundary
A——A′ **Line of cross section**
1○ **Well**—Number corresponds to table 2. Color indicates the hydrogeologic unit encountered at total depth

Figure 5. Surficial extent of hydrogeologic units and prominent structural geologic features of the Rush Valley area, Tooele County, Utah.—Continued

Table 2. Summary of selected drillers logs, Rush Valley, Tooele County, Utah.

[See figures 5 and 6 for well locations. All depths are in feet. Lithology encountered at the bottom of the hole: Q, unconsolidated basin fill; Tsl, semiconsolidated to consolidated Tertiary-age Salt Lake Formation; IPo, Oquirrh Group bedrock. —, no data]

Figure ID[1]	Well ID[2]	Cross section[1]	Projection[3]	Basin-fill thickness[4]	Total well depth	Lithology	Clay layers[5]	Well completion date
1	427256	A-A'	3,280, NNE	400+	400	Q	62–84, 232–265	7/5/1969
2	431761	A-A'	1,770, NNE	450+	450	Q	—	8/28/2008
3	12995	B-B'	755, S	197+	197	Q	90–197	5/13/1947
4	14491	B-B'	4,890, N	240+	240	Q	35–65	11/27/1996
5	15661	C-C'	2,430, N	295+	295	Q	—	10/7/1997
6	431760	C-C'	165, NNW	203+	203	Q	—	9/22/2008
7	17053	C-C'	885, NNW	362+	362	Q	8–100	2/6/1998
8	23927	C-C'	165, SSE	250+	250	Q	16–43	9/13/2001
9	22414	C-C'	195, NNW	240+	240	Q	—	7/25/2000
10	14421	C-C'	130, NNW	138+	138	Q	87–115	11/21/1996
11	34503	C-C'	2,020, NNW	535+	535	Q	0–49, 69–275, 280–495, 503–540	10/12/2005
12	33740	D-D'	1,530, WNW	400+	400	Q	—	3/28/2005
13	428965	D-D'	1,410, ESE	1,140	1,165	IPo	35–75, 150–192, 195–230, 390–640, 1,015–1,120	9/30/1967
14	16880	D-D'	920, W	634	800	IPo	293–483	5/19/1998
15	13086	D-D'	100, W	547+	547	Q	48–163, 204–259, 280–355, 399–537	2/1/1975
16	13087	D-D'	1,690, ESE	440	585	IPo	13–80, 83–155, 240–321	11/18/1994
17	13075	D-D'	755, SE	275	360	Tsl	0–24, 35–60, 205–275	6/11/1952
18	27069	D-D'	625, SE	262+	262	Q	8–38, 80–136, 178–218	4/23/2003
19	18903	E-E'	1,440, WNW	240+	240	Q	—	3/23/1999
20	33956	E-E'	1,940, WNW	40	120	Tsl	—	4/2/2005
21	34134	E-E'	1,890, WNW	160+	160	Q	—	4/25/2005
22	21207	E-E'	720, ESE	94	469	Tsl	—	12/20/1999
23	13090	E-E'	4,270, ESE	285	298	Tsl	—	4/30/1945

[1] Figure ID corresponds to wells shown on figures 5 and 6

[2] Well ID uniquely identifies the drillers' log in the Utah Department of Natural Resources, Division of Water Rights well-drilling database where drillers' logs are available online at http://waterrights.utah.gov/wrdb/WINlookup.asp

[3] Approximate distance, in feet, and direction that well was projected onto the cross section For example, 3,280, NNE indicates that the well was projected 3,280 feet toward the north-northeast onto the cross section

[4] Thickness of unconsolidated basin-fill deposits, in feet

[5] Depth intervals where continuous clay layers were indicated on the drillers' log Drillers' logs are available from the Utah Department of Natural Resources, Division of Water Rights (2010)

thick to the east of Clover and in the south near Lofgreen and is composed primarily of tuffaceous sandstone, calcareous sandstone, tephra, and siltstone, with lesser amounts of marl, limestone, mudstone, and claystone. The distribution of each of these lithologies within the LBFAU is complex and laterally discontinuous. Tephra and tuffaceous or calcareous sandstone are the most common lithologies found in surface exposures of the LBFAU. Few wells are completed in the LBFAU, even where it occurs near the surface, because the permeability of these deposits is generally low.

Bedrock

Consolidated bedrock in Rush Valley consists primarily of Precambrian to Paleozoic-age sedimentary rocks. These rocks underlie basin-fill units in the subsurface and crop out in mountain ranges and hills in and bordering Rush Valley. Bedrock is divisible into four relevant hydrogeologic units based on their general hydrologic properties, lithology, and spatial distribution: the UCAU Pennsylvanian to Permian-age Oquirrh Group consisting of interbedded limestone, sandstone, and quartzite; the USCU upper Mississippian to lower Pennsylvanian-age fine-grained, siliciclastic sedimentary rocks; the LCAU Middle Cambrian to Mississippian-age limestone, dolomite, sandstone, shale, and quartzite; and the NCCU Precambrian to Lower Cambrian-age quartzite, shale, and diamictite.

The Oquirrh Group rocks that make up the UCAU contain the primary bedrock aquifer in Rush Valley. Outcrops of these rocks are found in many of the mountain ranges surrounding

Rush Valley and likely lie directly beneath much of the basin fill (figs. 5 and 6). Oquirrh Group lithologies include in upward succession limestone, interbedded limestone and sandstone, and sandstone or quartzite. Permeability of this unit is assumed to be medium to high depending on localized development of secondary porosity. These rocks readily yield water where wells intersect fracture systems in the Vernon area, along the north and south flanks of South Mountain, and near the mouth of Soldier Canyon east of Stockton.

The USCU consists of upper Mississippian to lower Pennsylvanian-age, shaly, siliciclastic, fine-grained rocks that overlie the Lower Paleozoic carbonate rocks. The dominant geologic unit classified as USCU in the study area is the Manning Canyon Shale. Rocks of the USCU are fine grained and have low primary permeability. Because of their low susceptibility to dissolution or fracturing, the USCU also lacks significant secondary permeability.

The Paleozoic rocks that comprise the LCAU include a thick section of limestone, dolomite, sandstone, and shale. This unit includes Middle Cambrian through Mississippian-age rocks that consist primarily of limestone and dolomite. This unit is laterally extensive (fig. 5) and likely hydrologically interconnected over wide areas. Few wells are completed in the LCAU and little is known about its quality as a producing aquifer. Although permeability of this unit locally may be medium to high where solutional widening of fractures and stratigraphic discontinuities have enhanced secondary porosity, significant low-permeability rocks also exist in this unit. These include parts of the Humbug Formation, the Long Trail Shale Member of the Great Blue Limestone, the Kanosh Shale, and the Dell Phosphatic Shale Member of the Deseret Limestone. Where these rocks are steeply dipping or highly faulted causing low-permeability rocks to be juxtaposed such that they likely impede groundwater movement, they are grouped with the USCU.

The NCCU consists of Precambrian and Lower Cambrian-age quartzite, shale, and diamictite and is exposed south of Vernon in the Sheeprock Mountains and along the crest of the Stansbury Mountains (fig. 5). Permeability of these rocks is likely low compared to the overlying (carbonate) hydrogeologic units, but may be locally enhanced where these rocks are fractured.

The VU consists of various igneous rocks exposed primarily along the southern margin of the study area in the Sheeprock and West Tintic Mountains and at isolated locations in the Oquirrh and East Tintic Mountains (fig. 5). The VU is included as a hydrogeologic unit because it is found in outcrop at scales that could influence groundwater movement in parts of the study area. Because of its varied lithology, the VU may act as a conduit or barrier to flow relative to surrounding hydrogeologic units. No wells in the study area are completed in this unit.

Hydrogeologic Cross Sections

Five simplified cross sections were constructed to illustrate the orientation of the hydrogeologic units in the subsurface for selected parts of the Rush Valley groundwater basin (fig. 6). The cross sections are based on recent geologic mapping completed by the Utah Geological Survey and several ongoing 1:24,000-scale quadrangle geologic mapping projects funded by the USGS STATEMAP program (Kirby, 2010a, b, c, d, e; Clark and others, 2009, 2010). Selected drillers' logs available from the Utah Department of Natural Resources, Division of Water Rights (2010), and regional-scale gravity data (Everitt and Kaliser, 1980; Bankey and others, 1998; Pan-American Center for Earth and Environmental Studies, 2010) were used to constrain unit thicknesses. Well logs used to construct the cross sections are listed in table 2.

Cross section A–A' extends west to east from bedrock on South Mountain to bedrock on the Oquirrh Mountains (fig. 6.1). The central third of the cross section is drawn through the UBFAU near Stockton. Bedrock of the Oquirrh Group (UCAU) lies along much of the cross section. These rocks consist of nearly equal amounts of interbedded sandstone and limestone. Bedding in the Oquirrh Group is steeply north dipping and strikes parallel or subparallel to the line of section. On the basis of gravity data and available well logs, it is assumed that the UBFAU directly overlies the UCAU with no intervening LBFAU along the line of section. Gravity data indicate that the total thickness of the UBFAU is at least 450 ft and possibly much thicker near the center of the section. The UBFAU thins towards bedrock exposures at South Mountain and thinly mantles Oquirrh Group bedrock east of Stockton. Bedrock of the UCAU and unconsolidated basin fill of the UBFAU may be partially offset across a series of concealed down-to-the west normal faults near Stockton.

Cross section B–B' extends west to east from bedrock of the Oquirrh Group (UCAU) exposed in the eastern Stansbury Mountains to UCAU exposures at Indian Hill and a perennial spring at the eastern edge of the cross section (fig. 6.1). Eastward sloping alluvial fans comprising the UBFAU probably mantle bedrock east of the Stansbury Mountains. A pair of prominent fault scarps with opposing offset defines a small graben between Indian Hill and the Stansbury Mountains. On the basis of well logs, the total thickness of the UBFAU in the graben based on well logs is at least 240 ft but likely not much greater than 300 ft. Gravity data suggest that the UBFAU thinly mantles bedrock both east and west of Indian Hill. Bedding in the UCAU is steeply dipping or overturned and strikes perpendicular to the line of section west of the graben. Bedrock near Indian Hill is west dipping and nearly perpendicular to the line of section.

Cross section C–C' extends west to east across Johnson Pass, paralleling Clover Creek and State Highway 199 toward the center of Rush Valley (figs. 5 and 6.1). Gently folded and faulted Paleozoic bedrock of the LCAU is encountered along the western third of the cross section. The east-dipping UCAU overlies the LCAU between the projected location of Clover

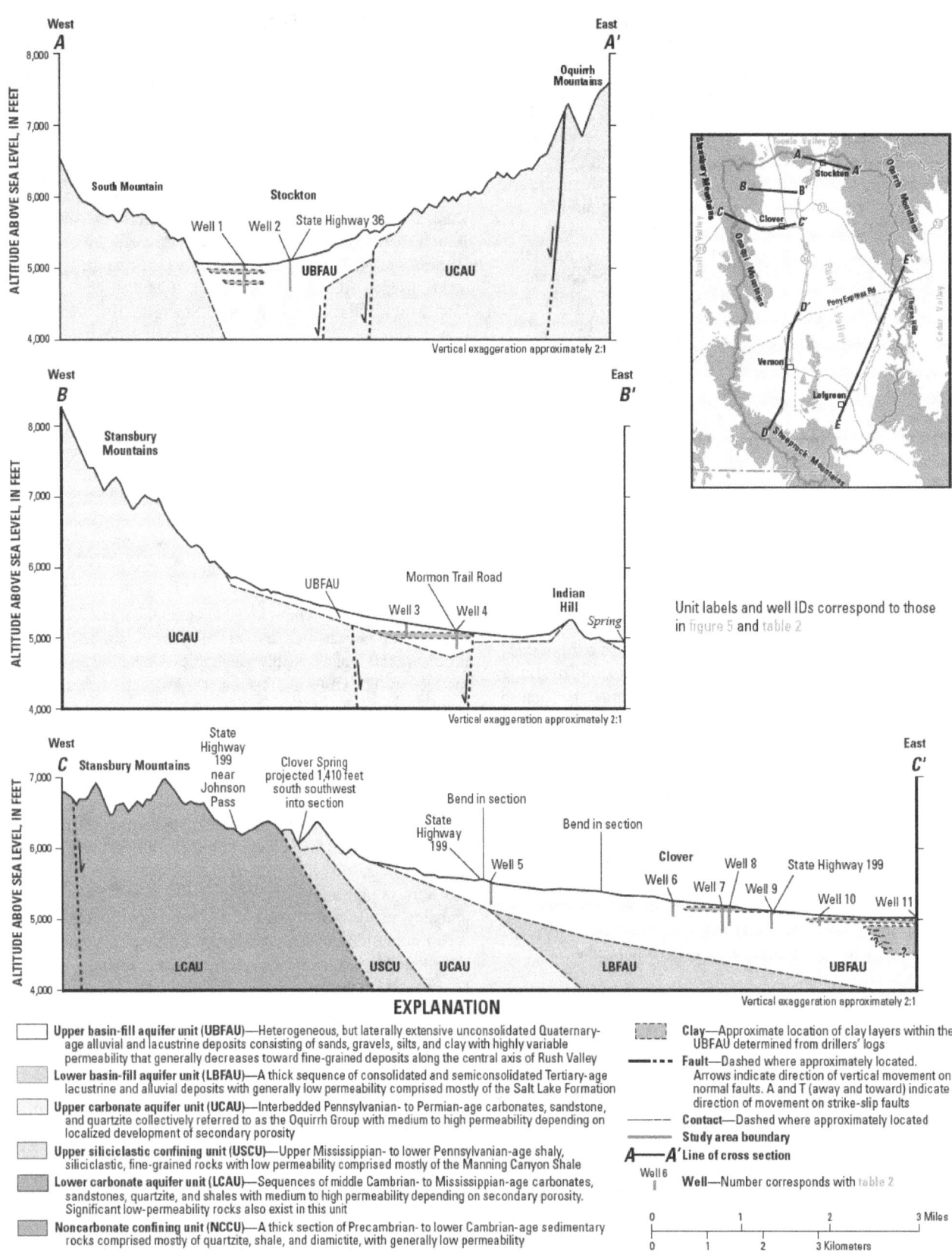

Unit labels and well IDs correspond to those in figure 5 and table 2

EXPLANATION

Upper basin-fill aquifer unit (UBFAU)—Heterogeneous, but laterally extensive unconsolidated Quaternary-age alluvial and lacustrine deposits consisting of sands, gravels, silts, and clay with highly variable permeability that generally decreases toward fine-grained deposits along the central axis of Rush Valley

Lower basin-fill aquifer unit (LBFAU)—A thick sequence of consolidated and semiconsolidated Tertiary-age lacustrine and alluvial deposits with generally low permeability comprised mostly of the Salt Lake Formation

Upper carbonate aquifer unit (UCAU)—Interbedded Pennsylvanian- to Permian-age carbonates, sandstone, and quartzite collectively referred to as the Oquirrh Group with medium to high permeability depending on localized development of secondary porosity

Upper siliciclastic confining unit (USCU)—Upper Mississippian- to lower Pennsylvanian-age shaly, siliciclastic, fine-grained rocks with low permeability comprised mostly of the Manning Canyon Shale

Lower carbonate aquifer unit (LCAU)—Sequences of middle Cambrian- to Mississippian-age carbonates, sandstones, quartzite, and shales with medium to high permeability depending on secondary porosity. Significant low-permeability rocks also exist in this unit

Noncarbonate confining unit (NCCU)—A thick section of Precambrian- to lower Cambrian-age sedimentary rocks comprised mostly of quartzite, shale, and diamictite, with generally low permeability

Clay—Approximate location of clay layers within the UBFAU determined from drillers' logs

Fault—Dashed where approximately located. Arrows indicate direction of vertical movement on normal faults. A and T (away and toward) indicate direction of movement on strike-slip faults

Contact—Dashed where approximately located

Study area boundary

A——A' Line of cross section

Well 6   Well—Number corresponds with table 2

0   1   2   3 Miles

0   1   2   3 Kilometers

Figure 6.   Schematic hydrogeologic cross sections (A–A', B–B', and C–C') for selected locations in Rush Valley, Tooele County, Utah.

EXPLANATION

Upper basin-fill aquifer unit (UBFAU)—Heterogeneous, but laterally extensive unconsolidated Quaternary-age alluvial and lacustrine deposits consisting of sands, gravels, silts, and clay with highly variable permeability that generally decreases toward fine-grained deposits along the central axis of Rush Valley

Lower basin-fill aquifer unit (LBFAU)—A thick sequence of consolidated and semiconsolidated Tertiary-age lacustrine and alluvial deposits with generally low permeability comprised mostly of the Salt Lake Formation

Upper carbonate aquifer unit (UCAU)—Interbedded Pennsylvanian- to Permian-age carbonates, sandstone, and quartzite collectively referred to as the Oquirrh Group with medium to high permeability depending on localized development of secondary porosity

Upper siliciclastic confining unit (USCU)—Upper Mississippian- to lower Pennsylvanian-age shaly, siliciclastic, fine-grained rocks with low permeability comprised mostly of the Manning Canyon Shale

Lower carbonate aquifer unit (LCAU)—Sequences of middle Cambrian- to Mississippian-age carbonates, sandstones, quartzite, and shales with medium to high permeability depending on secondary porosity. Significant low-permeability rocks also exist in this unit

Noncarbonate confining unit (NCCU)—A thick section of Precambrian- to lower Cambrian-age sedimentary rocks comprised mostly of quartzite, shale, and diamictite, with generally low permeability

Clay—Approximate location of clay layers within the UBFAU determined from drillers' logs

Fault—Dashed where approximately located. Arrows indicate direction of vertical movement on normal faults. A and T (away and toward) indicate direction of movement on strike-slip faults

Contact—Dashed where approximately located

Study area boundary

D——D' Line of cross section

Well 6 **Well**—Number corresponds with table 2

Unit labels and well IDs correspond to those in figure 5 and table 2

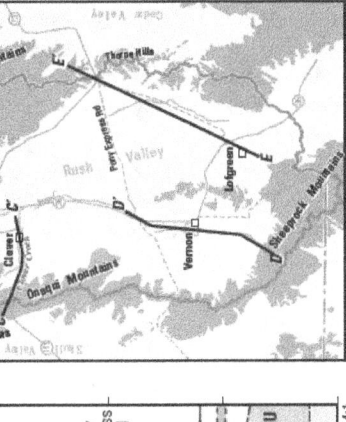
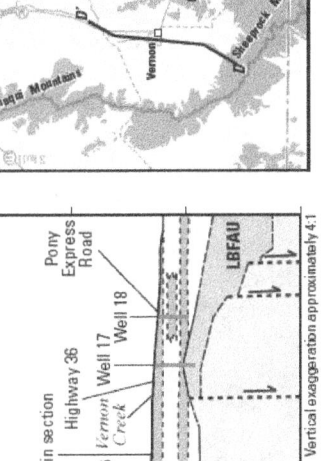
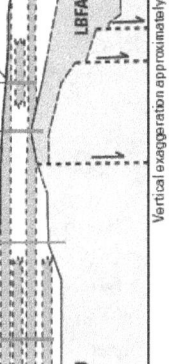
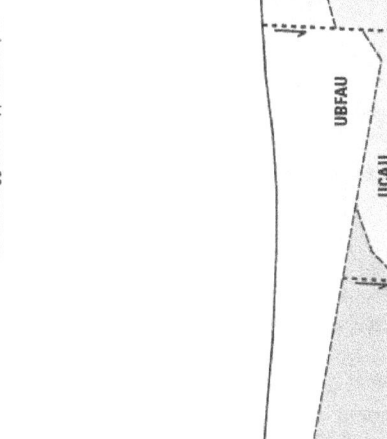

Figure 6. Schematic hydrogeologic cross sections (*D–D'* and *E–E'*) for selected locations in Rush Valley, Tooele County, Utah.—Continued

Spring and the UBFAU that covers bedrock across the eastern two-thirds of the cross section (fig. 6A). Total thickness of the UBFAU is unconstrained by well logs, but gravity data and correlation with areas of known thickness to the south indicate a dramatic eastward thickening across the cross section. Thickness of the UBFAU likely increases eastward to at least 500 to 600 ft. Beneath the UBFAU, the older LBFAU is inferred to account for most of the deep basin fill. Available data indicate that the LBFAU has a limited lateral extent in the subsurface, leaving a zone where the UBFAU directly overlies bedrock along the eastern flank of the southern Stansbury Mountains.

Cross section D–D' extends northward from the Sheeprock Mountains through the community of Vernon roughly parallel to State Highway 36 (figs. 5 and 6B). Precambrian-age rocks of the NCCU are shown in the southernmost part of the cross section. North-dipping rocks of the LCAU and the UCAU overlie the NCCU, with the Oquirrh Group of the UCAU forming the uppermost bedrock across most of the section. The UBFAU covers all bedrock north of the Sheeprock Mountains. The UBFAU gradually thickens north of the Sheeprock Mountains to the Vernon area, where the log of well 13 indicates just over 1,000 ft of unconsolidated deposits lying directly on the UCAU (table 2; figs. 5 and 6B). North of this well, a down-to-the south/southwest fault that may be the northwestward continuation of the Vernon Creek fault is inferred to offset bedrock. North of this structure, the UBFAU gradually thins northward from nearly 650 to 200 ft near the intersection of State Highway 36 and the Pony Express Road. Northward-thickening LBFAU overlies the UCAU along the north end of the cross section.

Cross section E–E' extends from south-southwest to north-northeast across southeastern Rush Valley through Lofgreen to the southern Oquirrh Mountains north of Fivemile Pass (figs. 5 and 6B). Well logs and intermittent exposures of the Salt Lake Formation indicate a thin mantle of UBFAU, less than 300 ft thick, overlies the Salt Lake Formation along much of this cross section. The UBFAU is assumed to thicken in the north across several basin-bounding normal faults and alluvial fans that slope southwestward from the Oquirrh Mountains. Underlying the UBFAU, the LBFAU is variable in thickness. The UCAU may underlie the Salt Lake Formation along most of the section, and the LCAU replaces the UCAU along the northernmost part of the section where it is exposed at the surface near Fivemile Pass.

Aquifer Properties

Aquifer properties describe the ability of a groundwater system to transmit and store water. The distribution of these properties in Rush Valley is variable and depends on the depositional environment of sediments in the basin-fill aquifer and on the degree of structural deformation, fracturing, and/or chemical dissolution in the bedrock aquifers. Aquifer properties can be estimated with aquifer tests by pumping groundwater from a well and monitoring the water-level changes in the pumped well or in nearby observation wells.

Because this method results in localized values that are generally representative of conditions near the pumped well, it may not represent the variability and heterogeneity throughout the aquifer.

Aquifer test data are commonly used to estimate values of transmissivity (T) and hydraulic conductivity (K). Both of these are measures that describe the ease with which water can move through the pore spaces or conduits within an aquifer. More specifically, K is the volume of water flowing through a unit cross-sectional area of an aquifer under a unit hydraulic gradient in a given amount of time, and T is the volume of water flowing through a cross-sectional area that is one unit wide multiplied by the aquifer thickness in a given amount of time. The quantities are proportionally related by the aquifer thickness as shown below:

$$T = Kb \qquad (1)$$

where:

T is the transmissivity (in ft^2/day),

K is the hydraulic conductivity (in ft/d), and

b is the aquifer thickness (in ft).

Well log data indicate that most wells in the Rush Valley study area produce groundwater from discrete layers or lenses of coarse-grained sediments in the basin fill and from fracture zones in bedrock. In basin fill, these coarse-grained zones may be the buried remnant of old fluvial channels. One example comes from a driller's log of a well located in T. 8 S., R. 5 W., section 30 along the Vernon Creek-Faust Creek trend that notes drilling was terminated at only 53 ft when gravel continued to fill the hole as fast as it could be removed. Another example is well (C–5–5)32dbb–2, drilled to a depth of 112 ft in sands and gravels along the Clover Creek drainage. This reportedly is one of the best producing wells in the Clover-St. John area (James Schlosser, oral commun.). The basin fill also contains thick zones of fine-grained, low-permeability deposits.

Data are not sufficient to accurately estimate the extent of these permeable zones in the bedrock or basin-fill aquifers. Furthermore, the contributing thickness (b) of the aquifer in equation 1 is rarely well known. Because of the uncertainty in b, aquifer-test data were used to estimate values of T rather than K in Rush Valley (table 3).

Basin Fill

Data from 16 single-well aquifer tests reported on drillers' logs were used to estimate the T of the basin fill. Values of T were determined by using the discharge/drawdown relationship from drillers' logs and AQTESOLV™ software for Windows to simplify a successive approximation method based on the Cooper-Jacob solution for flow to a well in a confined aquifer (Cooper and Jacob, 1946). To include as many wells as possible, the analyses were performed for wells that were pumped (rather than bailed) for a minimum of

Table 3. Aquifer properties determined from 25 single-well aquifer tests and one multiple-well aquifer test in Rush Valley, Tooele County, Utah.

[Location: See "numbering system" at beginning of report for an explanation of the numbering system used for hydrologic-data sites in Utah UBFAU, upper basin-fill aquifer unit; LBFAU, lower basin-fill aquifer unit; UCAU, upper carbonate aquifer unit; NR, not reported; —, no data]

Location	Date of test	Screened interval of well, in feet below land surface	Bedrock or basin fill and hydrogeologic unit	Pumping rate, in gallons per minute	Duration, in hours	Draw-down, in feet	Specific capacity, in gallons per minute per foot of drawdown	Transmissivity, in feet squared per day
(C-5-5)5adb-1	September 1973	51–168	Basin fill, UBFAU[1]	1,140	21	77	14.8	3,900[2]
(C-5-5)11bba-1	August 2006	211–236	Basin fill, UBFAU[3]	75	3	40	1.9	280[2]
(C-5-5)30dac-1	May 2008	39–79	Basin fill, UBFAU[1]	20	4	15	1.3	290[2]
(C-5-5)30ddb-1	December 2008	65–105	Basin fill, UBFAU[1]	12	30	60	0.2	50[2]
(C-5-6)25caa-1	November 2001	124–134, 242–247	Basin fill, UBFAU[3]	20	5	40	0.5	70[2]
(C-6-4) 4acc-1	November 1986	500–1,000	Basin fill, UBFAU[3]	700	17	70	10.0	1,800[2]
(C-6-7)3cba-1	May 2008	400–480	Basin fill, UBFAU[3]	20	6	100	0.2	30[2]
(C-7-5)16ada-1	August 2004	64–71	Basin fill, UBFAU[1]	82	1	25	3.3	700[2]
(C-7-5)28ccc-1	July 2002	260–300	Basin fill, UBFAU[1]	20	28	80	0.3	60[2]
(C-7-5)29dca-1	December 1995	142–250	Basin fill, UBFAU[1]	30	8	20	1.5	350[2]
(C-7-5)29ddb-1	April 2003	257–262	Basin fill, UBFAU[1]	30	30	2	15.0	4,400[2]
(C-8-5)20dcc-1	July 1978	137–270	Basin fill, UBFAU[3]	30	2	10	3.0	420[2]
(C-8-5)20dcc-1	July 1978	137–270	Basin fill, UBFAU[3]	60	8	26	2.3	370[2]
(C-8-6)12aca-1	April 2007	46–146	Basin fill, UBFAU[1]	30	20	100	0.3	60[2]
(C-8-6)25bab-1	February 1995	—	Basin fill, UBFAU[1]	15	8	200	0.1	10[2]
(C-10-4)14aab-1	September 2007	180–223	Basin fill, LBFAU[3]	10	10	12	0.8	140[2]
(C-4-4)32add-1	June 2008	420–620	Bedrock, UCAU[3]	1,120	24	60	18.7	3,700[2]
(C-4-5)8cbb-1	November 2008	NR	Bedrock, UCAU[4]	300	60	10	30.0	7,100–9,200[2]
(C-4-5)29bdc-2	August 2007	500–700	Bedrock, UCAU[4]	1,200	24	17	70.6	15,600–20,600[2]
(C-4-5)30aac-2	December 2002	545–705	Bedrock, UCAU[4]	600	24	17	35.9	7,200–9,700[2]
(C-8-3)3cbd-1	January 2007	430–498	Bedrock, LCAU[3]	147	72	159	0.9	170[2]
(C-8-3)10bcb-1	September 2007	1,087–1,347, 1,389–1,409, 1,451–1,551, 1,593–1,613	Bedrock, UCAU[4]	475	120	122	3.9	800–1,100[2]
(C-8-5)6ddb-2	November 1994	375–582	Bedrock, UCAU and UBFAU[1]	5,600	216	—	—	400,000[5]
(C-8-5)17ccc-1	May 1998	634–800	Bedrock, UCAU and UBFAU[1]	1,600	8	278	5.8	1,400[2]
(C-8-5)17ccc-1	May 1998	634–800	Bedrock, UCAU[1]	1,830	10	278	6.6	1,600[2]

[1] Aquifer confined, storativity of 0 001 assumed for Cooper-Jacob analysis

[2] Determined using method described by Cooper and Jacob (1946)

[3] Aquifer unconfined, storativity of 0 075 assumed for Cooper-Jacob analysis

[4] Aquifer condition uncertain, range of transmissivity presented using storativity values for unconfined (0 075) and confined (0 001) aquifers

[5] Determined using data from a 9-day multiple-well aquifer test and the method described by Theis (Domenico and Schwartz, 1998)

1 hour. This method assumes that the aquifer is homogeneous, isotropic, and infinite in extent; that wells are fully penetrating, that flow to the well is horizontal, and that water is released instantaneously from storage. Additionally, for unconfined aquifers, drawdown is assumed to be small relative to the saturated thickness of the aquifer. Because these assumptions are not explicitly met, the uncertainty associated with individual values of T may be significant. The Cooper-Jacob method requires that a value for storativity (S) be used to obtain an estimate of T. Values for S were assumed to be 0.075 and 0.001 for unconfined and confined aquifers, respectively. Calculated values of T were insensitive to assumed values of S, varying less than about 20 percent for a corresponding order of magnitude change in S. The aquifer-test data from wells in basin fill resulted in T values ranging from 10 to 4,400 ft²/d (table 3), with a median value of about 290 ft²/d. Eleven out of sixteen of these tests yielded estimates of T between 100 and 1,000 ft²/d, indicating that the median value of 290 ft²/d is representative of many of the wells screened in Rush Valley basin fill. Examination of the T estimates did not reveal any clear and systematic patterns of spatial variation, and values of T differ by more than two orders of magnitude over a distance as small as 2 mi in the Vernon area. Drillers' logs indicate that only one of the 16 basin-fill wells used to estimate T is screened in the LBFAU hydrogeologic unit. The remaining 15 wells are screened in the UBFAU.

Bedrock

Eight bedrock T values were determined using single-well discharge/drawdown data from drillers' logs and analyzed by the Cooper-Jacob method as described above. One value was determined using combined drawdown and recovery analysis from a 9-day multiple-well aquifer test performed in 1995 on well (C–8–5)6ddb–2 near Vernon. The aquifer test was analyzed by the USGS Utah Water Science Center in May 1995 to determine the T and S properties of the basin fill near Vernon and to determine if pumping well (C–8–5)6ddb–2 could affect water levels and discharge of the Vernon city wells located about 5 mi to the south. The unpublished results of the analysis were reviewed and approved and are contained in the files of the Utah Water Science Center.

Aquifer-test data from wells in bedrock resulted in T values ranging from 170 to 400,000 ft²/d (table 3) and a median value of 3,700 ft²/d. The higher values are likely representative of T only where carbonate rock is highly fractured. According to drillers' logs in the Utah Division of Water Rights well-drilling database (http://www.waterrights.utah.gov/cgi-bin/wellview.exe), attempts to drill productive wells in UCAU or LCAU bedrock occasionally result in abandonment because they yield little to no water. Examples are in Ophir Canyon (T. 5 S., R. 4 W., sections 13, 24, and 27) where several dry holes were drilled to depths of between 200 and 1,000 ft immediately adjacent to a perennial stretch of Ophir Creek, and in the Thorp Hills (T. 8 S., R. 3 W., section 5) where a driller's log reported 920 ft drawdown in response to a 6 gal/min pumping rate in a 920-ft deep test well.

Six of the nine aquifer tests report T values between 1,000 and 10,000 ft²/d. The multiple-well aquifer test yielded the highest value of T (400,000 ft²/d) within the study area. Considering that typical values of K for fractured limestone range from less than 1 to around 1,000 ft/d (Freeze and Cherry, 1979; Domenico and Schwartz, 1998), and assuming that a reasonable thickness for a productive aquifer is several hundred feet, this value of T is near the upper end of the expected range of values. The pumped well for this test, (C–8–5) 6ddb–2, is screened through both boulders and fractured bedrock. It is, therefore, unlikely that this high T can be attributed entirely to bedrock, because water was likely withdrawan from both high-permeability unconsolidated deposits and the underlying fractured rock of the UCAU. Several deep wells north of Vernon ((C–8–5)17ccc–1, (C–8–5)6ccd–1, and (C–8–5)7ddd–2) along the Vernon Creek-Faust Creek trend have encountered permeable gravels or fractured UCAU bedrock at depths of 550 to 800 ft. These wells yield thousands of gallons per minute throughout much of the irrigation season, indicating that this is one of the most reliable zones of high transmissivity in Rush Valley.

Occurrence and Movement of Groundwater

Groundwater in Rush Valley occurs in both unconsolidated basin-fill and in consolidated-rock aquifers under confined and unconfined conditions. Within the basin fill, unconfined or water-table conditions generally exist along the valley margins within alluvial fan and colluvial deposits, and confined conditions generally exist in the central parts of the valley. Groundwater moves under confined conditions where lacustrine and fluvial deposits have created zones of permeable material mixed with semicontinuous to continuous layers of low-permeability clay or silt. Although unconfined groundwater movement occurs within most bedrock mountain areas, structural geologic features and variations in lithology likely result in localized areas of confined conditions.

Groundwater generally moves from high-altitude recharge areas toward low-altitude discharge areas. As described by Hood and others (1969), groundwater in Rush Valley moves toward two different discharge areas. A water-level surface map, developed from water-level measurements made at more than 100 wells during the fall of 2008, shows general directions of groundwater movement (fig. 7, table 1–1). The location of the groundwater divide described by Hood and others (1969, pl. 1), extending from the eastern edge of the Onaqui Mountains northeastward across the valley to near the mouth of Ophir Canyon, appears essentially unchanged since that time. North of the divide, groundwater moves from the Stansbury, Onaqui, and Oquirrh Mountains surrounding the valley toward discharge areas along the valley's central axis and in the vicinity of Rush Lake and then northward toward Tooele Valley in the vicinity of the Stockton Bar. South of the divide, most groundwater moves northeastward from recharge areas in the Onaqui, Sheeprock, and West Tintic Mountains across the valley toward the vicinity of Fivemile

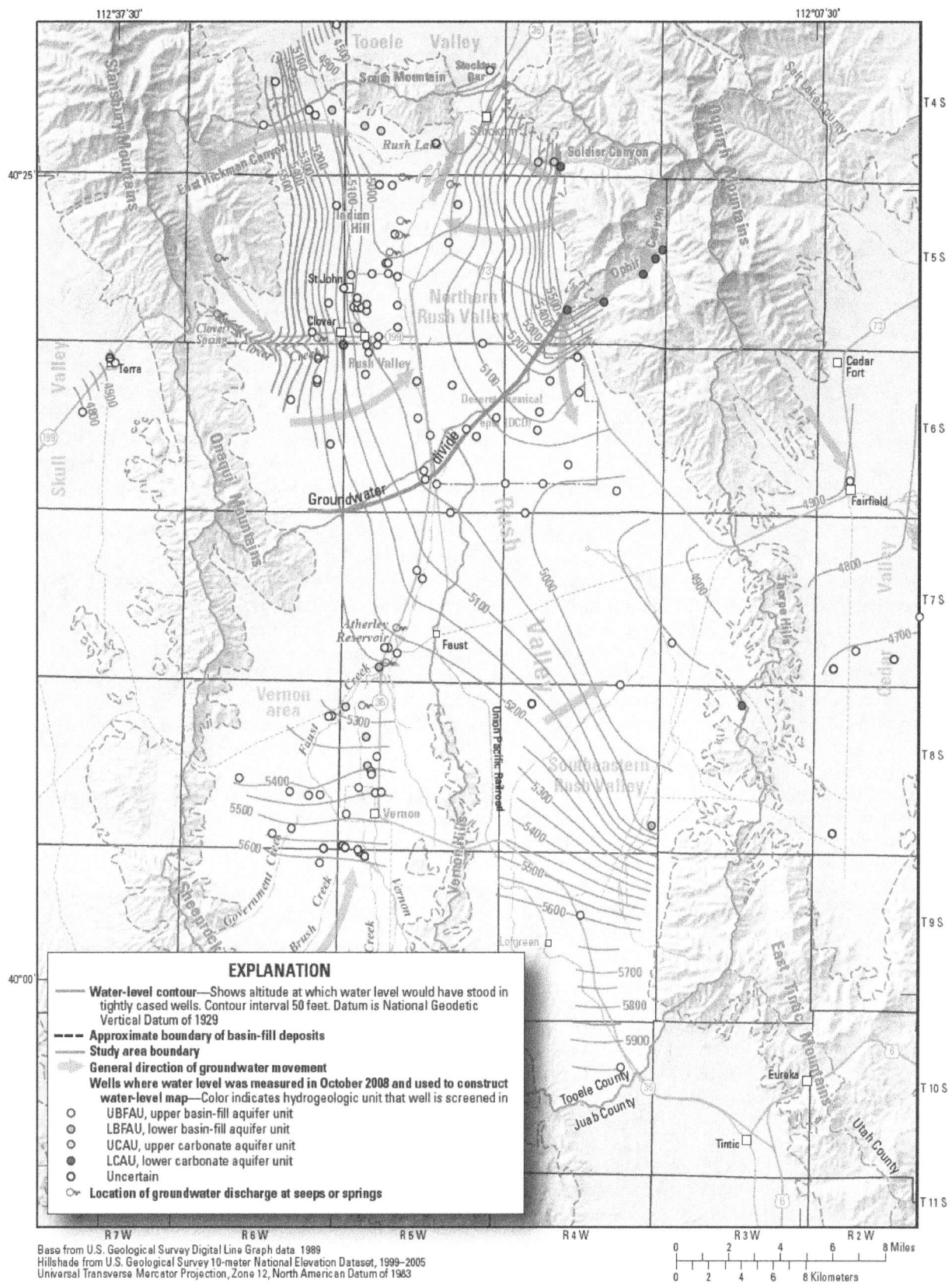

Figure 7. Regional water-level surface map for the Rush Valley study area, October 2008, Tooele County, Utah.

Pass at the southern end of the Oquirrh Mountains. A fraction of the groundwater originating south of Vernon moves to a significant area of discharge between Vernon and Faust. Some groundwater also moves southeastward along the front of the Oquirrh Mountains south of Ophir Canyon.

Although the water-level surface map indicates that groundwater is moving toward Tooele Valley at the Stockton Bar and toward Cedar Valley near Fivemile Pass, the combination of fine-grained basin fill and a nearly flat hydraulic gradient in the low-altitude parts of the valley indicate that the amount of interbasin flow is small and the rate of movement is slow. Groundwater-flow in each of the three areas of Rush Valley is discussed in the following sections.

Northern Rush Valley

Directly north of the groundwater divide in Rush Valley, groundwater moves eastward from recharge areas in the northern Onaqui and southern Stansbury Mountains and westward from the southern Oquirrh Mountains toward the center of the valley (fig. 7). Near the divide, groundwater moves through the UBFAU under generally unconfined conditions. The UBFAU reaches a thickness of at least 500 to 600 ft, and possibly as much as 1,000 ft, east of Clover along the boundary between T.5 S. and T.6 S., and it thins to the north where it directly overlies the UCAU near Indian Hill (cross section B–B', fig. 6A). Groundwater movement through the UCAU is likely impeded by the near vertically dipping bedding oriented perpendicular to the direction of groundwater movement where it underlies the UBFAU throughout this area. Assuming most groundwater moves through the UBFAU, the active groundwater-flow system in northern Rush Valley is likely constrained to within about 1,000 ft of the land surface.

As the groundwater-flow paths converge near the valley center, groundwater is directed northward across a discharge area that extends past Indian Hill to near the Stockton Bar. Groundwater conditions are confined north of about the T.5 S. and T.6 S. boundary in the valley center where discharge occurs at springs and flowing wells. Because Rush Valley is a closed basin with regard to surface water, all water that discharges at springs or flowing wells is ultimately consumed by ET. The gentle slope of the water-level surface and low permeability of aquifer material in the center of northern Rush Valley indicate that groundwater is moving slowly through this area and likely discharges to ET, springs, or flowing wells. Groundwater not consumed by ET continues to move toward the lowest altitude in Rush Valley at the northern end of Rush Lake and toward Tooele Valley. Water-level altitudes in three of four wells penetrating the UCAU along the south slope of South Mountain [(C–4–5)27cdb–1, 29bdc–2, and 30aac–1)] are about 140 to 240 ft lower (table A1–1) than in the neighboring UBFAU. Although this indicates the potential for groundwater to move from the UBFAU into the UCAU of South Mountain, the complicated geologic structure of South Mountain likely prevents flow in this direction and may only allow flow eastward around the bedrock block and through the UBFAU toward Tooele Valley at the Stockton Bar (fig. 7).

Vernon Area

South of the groundwater divide in the Vernon area, groundwater moves north from the recharge area in the Sheeprock Mountains toward a large area of discharge that extends from south of Vernon (near the boundary between T. 8 S. and T. 9 S.) northward to the vicinity of Atherley Reservoir (fig. 7). Groundwater conditions in the UBFAU are unconfined in the southern and westernmost parts of the valley near the basin-fill boundary and transition to confined conditions near the discharge area. The UBFAU thickens north of the Sheeprock Mountains to just over 1,000 ft near Vernon where it directly overlies the UCAU (cross section D–D', fig. 6B). From this point, the UBFAU thins northward, first abruptly to around 650 ft, then more gradually to around 200 ft near the intersection of State Highway 36 and the Pony Express Road. This northward thinning of the UBFAU is likely responsible for discharge in the Vernon area as groundwater is forced upward by the underlying low-permeability parts of UCAU and semiconsolidated LBFAU. Discharge in this area occurs by ET, to focused and diffuse springs and flowing wells, and as baseflow to Faust Creek. Groundwater not consumed by ET continues to move northward past the Vernon Hills and then eastward toward Cedar Valley at Fivemile Pass.

A few high-yield wells produce water from the UCAU north of Vernon. Drillers' logs from five wells [(C–8–5)6ccd–1, 6ddb–1, 6ddb–2, 7ddd–2, and 17ccc–1] indicate the presence of a highly fractured UCAU zone directly underlying a 20-to-100-ft thick zone of highly permeable quartzite cobbles and boulders. The UCAU is confined in this area by thick sequences of clay in the UBFAU. The quartzite cobbles and boulders may be fluvial deposits from the Sheeprock Mountains to the south. Water-level altitudes (October 2009) in wells (C–8–5)6ccd–1, 7ddd–2, and 17ccc–1 are approximately 5,241–5,246 ft, (table A1–1), indicating a very low hydraulic gradient in this deep aquifer over a minimum distance of at least 2 mi north to south and 1 mi east to west (fig. 7). Although these water-level altitudes are higher than the altitude of the clay confining units in the overlying UBFAU, signifying nonflowing artesian conditions, they are about 70 to 165 ft below water levels measured during October 2009 in wells finished in the overlying UBFAU.

Southeastern Rush Valley

Groundwater in the southern part of southeastern Rush Valley moves in a north-to-northeast direction from a relatively small recharge area in the West Tintic Mountains (fig. 7). In the southernmost part of this sub area, the UBFAU occurs only as a thin mantle overlying the low-permeability LBFAU and is assumed to thicken toward the north-northeast (cross section E–E', fig. 6B). Unconfined conditions probably exist throughout most of southeastern Rush Valley. Water-level contours indicate that no significant recharge moves into the valley from the East Tintic Mountains or from the hills extending north of them. This is likely due to low precipitation rates combined with geologic structures that prevent westward groundwater movement out of these highlands, where the bedrock is extremely faulted and folded (fig. 5). In the northern part of southeastern Rush Valley, groundwater moves south-southeasterly away from the valley groundwater divide, paralleling the southern Oquirrh Mountain range front. Water-level contours in this area also indicate that no significant recharge moves into the valley from the neighboring Oquirrh Mountains. Hood and others (1969) and Feltis (1967) noted an area in the southern Oquirrh Mountains within the Rush Valley surface-water drainage where geologic structural features direct groundwater recharge toward Cedar Valley. This is described in more detail in the "Recharge" subsection of this report. No discharge occurs at the surface in southeastern Rush Valley, where water levels are more than 120 ft below land surface near the valley bottom. Small amounts of groundwater from the Vernon area converge with the northerly and southerly flow paths and move toward Cedar Valley.

Groundwater Budget

The groundwater budget presented here is compiled from estimates of average annual recharge to and discharge from the Rush Valley groundwater basin. Although records used to construct the individual budget components vary somewhat in their temporal length, the budget is intended to represent average conditions over approximately the last half century. Previous studies of western basins in Utah generally developed groundwater budgets that focused only on the alluvial (valley) part of the basin (Hood and others, 1969) where groundwater was being developed as a resource.

In recent years, groundwater development has targeted fractured bedrock beneath the alluvium and at the base of the surrounding mountains. For this reason, the groundwater budget compiled for Rush Valley uses annual net recharge and discharge of the complete groundwater system—including the bedrock aquifers of the surrounding mountains. Because the groundwater budget encompasses the interconnected mountain and valley aquifers, it must account for intermediate forms of recharge and discharge. The following discussion provides an explanation of tracking the intermediate budget components.

Infiltration of mountain precipitation is the component of recharge that enters the groundwater system at the highest altitude. Some of this recharge moves through shallow parts of the mountain aquifers, discharges in drainage bottoms, and becomes baseflow to mountain streams and springs. The remaining fraction moves deeper through the mountain bedrock and into the adjoining basin-fill aquifer through the subsurface. Thus, the annual baseflow (table 1) of the mountain streams was included in the discharge portion of the budget (table 4) to reflect that not all of the recharge from infiltrating precipitation directly replenishes the downgradient basin-fill aquifer. Average annual flow in perennial mountain streams includes both a baseflow and runoff component, as discussed earlier in the streamflow section of this report. The baseflow represents groundwater discharging to streams in the mountains. Essentially all of the flow in these streams (baseflow and runoff combined) is captured and redistributed for use in the valley. Most is used for irrigation in areas where a fraction of the irrigation water is lost to seepage that, in turn, becomes aquifer recharge. A percentage of the total annual streamflow, therefore, is included in the recharge portion of the budget as "infiltration of unconsumed irrigation water."

Table 4.   Average annual groundwater budget by geographic area and for all of Rush Valley, Tooele County, Utah.

[Average annual volume in acre-feet per year]

Budget component	Northern Rush Valley	Vernon area	Southeastern Rush Valley	Valley total, rounded	Uncertainty[1] (in percent)	Percentage of total
Recharge						
[2]Infiltration of precipitation	18,600	7,400	5,900	**32,000**	50	82
[3]Infiltration of unconsumed irrigation water	3,200	1,600	2,100	**6,900**	20	18
Total recharge (rounded)	21,800	9,000	8,000	**39,000**	45	100
Discharge						
Evapotranspiration of groundwater	13,900	10,200	1,100	**25,000**	30	58
[4]Discharge to mountain springs and stream baseflow	4,700	4,200	2,300	**11,000**	60	26
[5]Well discharge	610	5,300	270	**6,200**	30	14
Subsurface discharge to Cedar and Tooele Valleys	600	0	300	**900**	170	2
Total discharge (rounded)	19,800	19,700	4,000	**43,000**	40	100

[1]Uncertainty estimates are explained in Appendix 2

[2]Recharge from the infiltration of precipitation was estimated using the Basin Characterization Model (1940–2006 average annual in-place recharge)

[3]Recharge from infiltration of unconsumed irrigation water is estimated to be 30 percent of the gaged or estimated streamflow from table 1

[4]Discharge to mountain springs and stream baseflow is derived from the estimated baseflow for each perennial mountain stream from table 1

[5]The division of well discharge between geographic areas in Rush Valley is based on 91 percent of irrigation pumping occurring in the Vernon area and the remaining 9 percent occurring in northern Rush Valley according to measurements of discharge from irrigation wells made by the U S Geological Survey Also, it is assumed that all industrial pumping (170 acre-feet) occurs at the Deseret Chemical Depot and that pumping from public supply, domestic, and stock wells (320 acre-feet) is divided evenly between the three geographic areas

The groundwater budget presented in table 4 lists budget components subdivided by the three geographic areas of the Rush Valley groundwater-flow system and for Rush Valley in its entirety. Average annual recharge to and discharge from all of Rush Valley are estimated to be 39,000 and 43,000 acre-ft, respectively. The majority of recharge and discharge occurs in northern Rush Valley, and less groundwater circulates through the Vernon area and southeastern Rush Valley. In the Vernon area budget, discharge appears to exceed recharge by more than 10,000 acre-ft. Some of this excess may be due to the withdrawal of groundwater from storage in the UCAU that is used for irrigation. It is also possible that the infiltration of precipitation is too low in the Vernon area. This is because the location of the groundwater divide in the Onaquai Mountains used to partition mountain recharge between northern Rush Valley and the Vernon area to the south is not well known. In the budget for southeastern Rush Valley, recharge appears to exceed discharge by about 4,000 acre-ft. Most of this excess is from the infiltration of precipitation that occurs in a high-altitude part of Ophir Canyon (fig. 8). The infiltration of precipitation may be too high in this area because, as in the Onaquai Mountains, the location of the groundwater divide used to partition mountain recharge between northern Rush Valley and southeastern Rush Valley in the Oquirrh Mountains is not well known.

The valley-wide budget shows a deficit in recharge of about 4,000 acre-ft/yr (table 4). Although this difference may reflect a basin-wide imbalance that indicates groundwater is being removed from storage, it is also well within the uncertainty that is inherent in the individual budget estimates. Groundwater budget uncertainty is discussed in more detail in Appendix 2.

Recharge

Recharge to the Rush Valley groundwater system is almost entirely by direct infiltration of snowmelt and rainfall that occurs in the mountains surrounding Rush Valley. The amount and distribution of recharge controls water levels and groundwater movement throughout much of the alluvial basin. A small but significant amount of recharge also occurs from the infiltration of unconsumed irrigation water sourced from mountain streams and springs.

Infiltration of Precipitation

The average annual recharge to the Rush Valley groundwater system from the infiltration of precipitation is estimated to be about 32,000 acre-ft (table 4). Direct or in-place groundwater recharge from precipitation was determined using the BCM. The BCM is a distributed-parameter water-balance accounting model used to identify areas having climatic and geologic conditions that allow precipitation to become runoff or in-place recharge and to estimate the amount of each (Flint and Flint, 2007, p. 6). The BCM calculations were made on a 270-m grid for each water year from 1940 to 2006. This 67-year period was selected because it encompassed the most up-to-date

BCM recharge and runoff estimates available for this part of the Great Basin and because limited climatic data are available prior to the 1940s. In-place recharge is calculated as the volume of water per time that percolates through the soil zone past the root zone and becomes net infiltration to consolidated rock or unconsolidated deposits. Runoff is the volume of water per time that runs off the surface. Runoff may infiltrate the subsurface farther downslope, undergo ET, or become streamflow. The BCM does not track or route runoff to determine its fate. The groundwater budget presented in this report assumes that BCM runoff is either lost to ET or becomes streamflow. Because essentially all streamflow in Rush Valley is captured and delivered for irrigation, the fraction of runoff that infiltrates farther downslope is accounted for in the 30 percent of water applied for irrigation that becomes recharge as discussed below. The BCM estimates of groundwater recharge and runoff approximated for subareas of the groundwater basin (fig. 8) are listed in table 5.

An advantage of using a distributed-parameter water-balance model, such as the BCM, is that the model identifies likely locations where runoff and in-place recharge are generated based on the temporal and spatial distribution of precipitation, snowmelt, sublimation, ET, soil-storage capacity, and saturated hydraulic conductivity. The spatial variation of in-place recharge is controlled more by altitude and geology than by other parameters. The highest rates generally occur at the highest mountain altitudes, where precipitation is greatest, and in areas where soil and bedrock are permeable. Average (1940–2006) in-place recharge is as much as 27 in./yr in the Oquirrh Mountains and as much as 10 to 11 in./yr in the Sheeprock and Stansbury Mountains, respectively (fig. 8). Although maximum precipitation rates are considerably lower in the Sheeprock Mountains than in the Stansbury Mountains, maximum recharge rates are similar. The BCM simulates very little recharge at valley altitudes. This is in agreement with Hood and others (1969) who noted that the amount infiltrating precipitation is generally small and most is held as soil moisture before being subsequently lost to ET.

During 1940–2006, annual BCM recharge ranged from 11,100 to 66,700 acre-ft (fig. 9). Compared to precipitation, in-place recharge has larger annual variations. It is higher during very wet years and greatly diminished during very dry years (Gates, 2007; Masbruch and others, 2011, fig. D–3), mainly due to ET in the recharge areas. During wet periods more water is available than is consumed by vegetation and during dry periods, vegetation generally maintains its usual rate of ET. As a result there is more groundwater recharge during wet periods and less during dry periods than would be estimated from a simple ratio of recharge to average annual precipitation.

Hood and others (1969) state that (1) most mountain recharge areas surrounding Rush Valley are composed of folded and fractured carbonate rocks that dip steeply toward the valleys and readily transmit water; (2) the maximum precipitation rate in the Sheeprock Mountains is less than in the other high ranges, but that the lithology and geologic structure of the rocks in that area aid in delivery of water to the

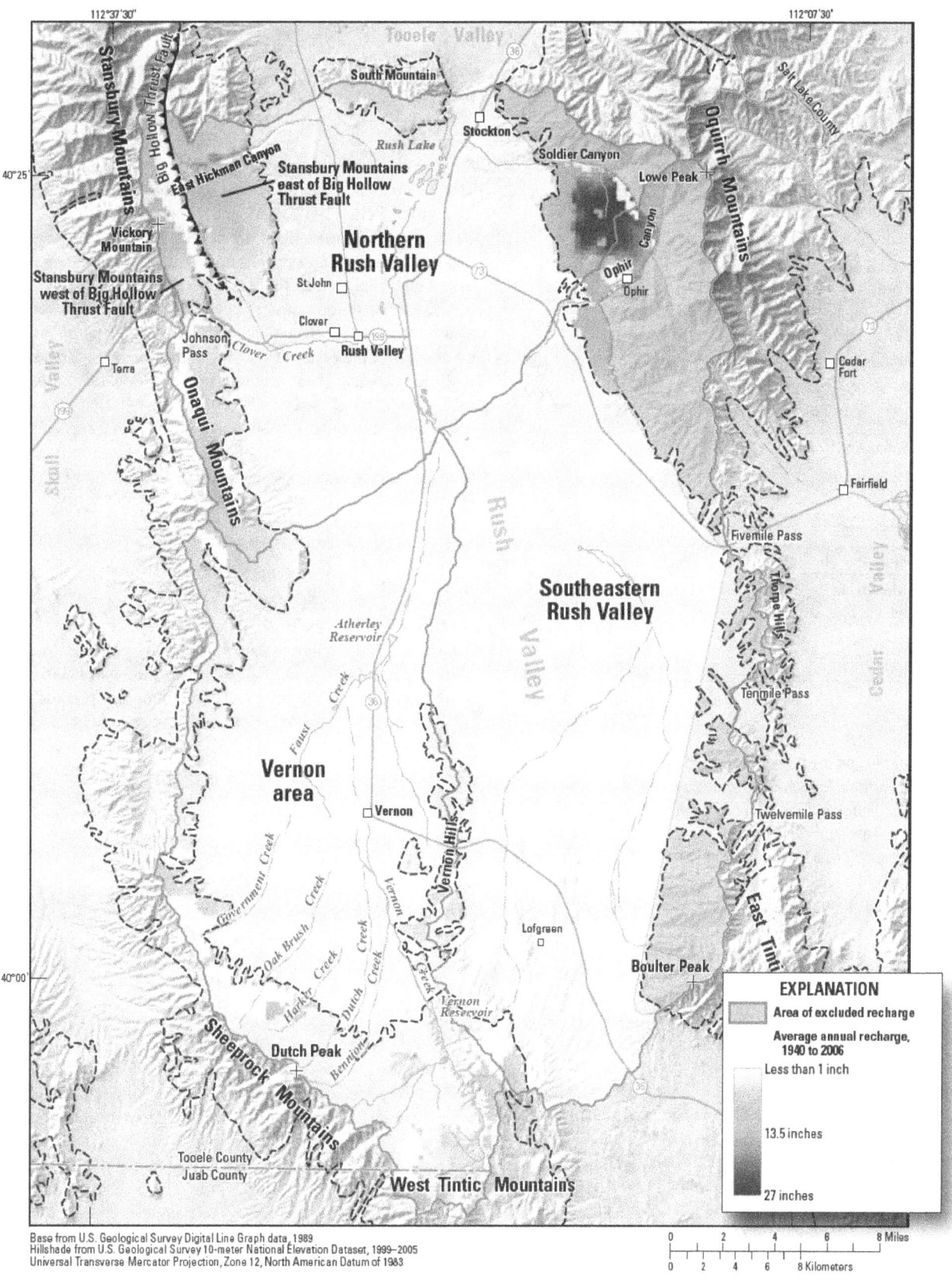

Figure 8.   Average annual (1940–2006) recharge from precipitation, Rush Valley, Tooele County, Utah.

Table 5.   Average annual in-place groundwater recharge and runoff from precipitation estimated by the Basin Characterization Model for 1940–2006 for Rush Valley, Tooele County, Utah.

[BCM, Basin Characterization Model; recharge and runoff values in acre-feet per year All components rounded to the nearest 100]

Area	BCM in-place recharge	BCM runoff
Northern Rush Valley		
Oquirrh Mountains	8,600	3,600
South Mountain	700	40
Stansbury Mountains, east of Big Hollow Thrust Fault	4,400	600
Stansbury Mountains, west of Big Hollow Thrust Fault	2,200	1,400
Onaqui Mountains	2,700	1,000
Total	**18,600**	**6,600**
Vernon area		
Onaqui Mountains	1,800	300
Sheeprock Mountains	5,000	13,100
Vernon Hills	600	200
Total	**7,400**	**13,600**
Southeast Rush Valley		
Oquirrh Mountains	5,100	1,800
West Tintic Mountains	400	800
Vernon Hills	400	200
Total	**5,900**	**2,800**
Total	**31,900**	**23,000**

valley; and (3) an area of the southern Oquirrh Mountains was excluded from their recharge estimate because the geologic structure is believed to inhibit recharge to Rush Valley. Findings of this study led to an expansion of this excluded area in the Oquirrh Mountains south of Ophir Creek and from the Thorpe Hills south through the East Tintic Mountains (fig. 8). Water-level contours are perpendicular to the range front in these areas, indicating that the higher-altitude water levels beneath these ranges, resulting from higher precipitation and infiltration rates, are not hydraulically connected to the valley groundwater system (fig. 7). The lack of recharge reaching southeastern Rush Valley south of the Thorpe Hills is explained by the limited precipitation (fig. 2) and the complex faulting in the areas of exposed bedrock (fig. 5). Structural controls on groundwater movement in the southern Oquirrh Mountains are supported by updated geologic mapping that shows steeply southwest-dipping beds of low-permeability shale units (the Manning Canyon and Long Trail Shale) of the USCU and LCAU parallel to the mountain front at or near the mountain-valley boundary south of Ophir Canyon (Clark and others, 2009). The orientation of these southwest-dipping beds is the result of a northwest-southeast-trending anticline in the southern Oquirrh Mountains and may contribute to mountain recharge being directed away from Rush Valley and toward Cedar Valley in this area.

Unconsumed Irrigation Water

Recharge from seepage of unconsumed irrigation water occurs where surface water, captured in the canyons of all perennial mountain streams, is distributed to agricultural areas outside of the discharge area in Rush Valley. It has been reported that between about 10 and 50 percent of water used for irrigation in similar climatic and hydrologic settings is not consumed by crops and becomes recharge to the basin-fill aquifer (Feltis, 1967, Clark and Appel, 1985; Stolp, 1994; Susong, 1995). The variation in the fraction of irrigation water that becomes recharge depends on factors such as the type of irrigation (flood, line sprinkler, center pivot, etc.) and local soil properties. This groundwater budget assumes that 30 percent of water applied for irrigation becomes recharge, similar to other areas of the eastern Great Basin that are highly irrigated with surface water (Heilweil and Brooks, 2011). It was also assumed that all of the estimated streamflow (table 1) is applied for irrigation. A small amount (less than 2 percent) of this water is captured from mountains streams and springs and used for municipal supplies in Stockton and in Ophir Canyon. However, accounting for this does not noticeably alter the basin-wide groundwater budget.

Discharge

Discharge from the Rush Valley groundwater system occurs by ET, discharge to mountain springs and baseflow to mountain streams, well withdrawals, and subsurface outflow to Tooele and Cedar Valleys. All ET of groundwater presented in the Rush Valley groundwater budget is assumed to occur in the valley bottoms; groundwater ET in the mountains is included in the calculation of BCM net infiltration. Much of the groundwater that discharges from significant mountain springs or as baseflow to mountain streams is captured and diverted to be used for public supply or irrigation. Groundwater that discharges from springs in the valley is eventually consumed by ET and is therefore included in the estimate for that component of the groundwater budget. Groundwater withdrawn by wells is used for irrigation, industrial use, public and domestic supply, and stock watering.

Evapotranspiration

Discharge of groundwater by ET is the combination of groundwater consumed by plants with roots that extend to the shallow water table and direct evaporation from areas of open water or soils that are wetted by shallow groundwater. The total volume of water discharged by ET can be calculated as the product of the rate at which water is transferred from the land to the atmosphere (ET rate, in ft/yr) and the acreage of the vegetation, open water, and soils that transfer this water. Groundwater ET, the fraction of total ET made up of groundwater and referred to herein as ETg, is calculated by subtracting precipitation and delivered irrigation water from the total ET. Average annual ETg in all of Rush Valley is estimated to be about 25,000 acre-ft/yr, with about 14,000 acre-ft/yr occur-

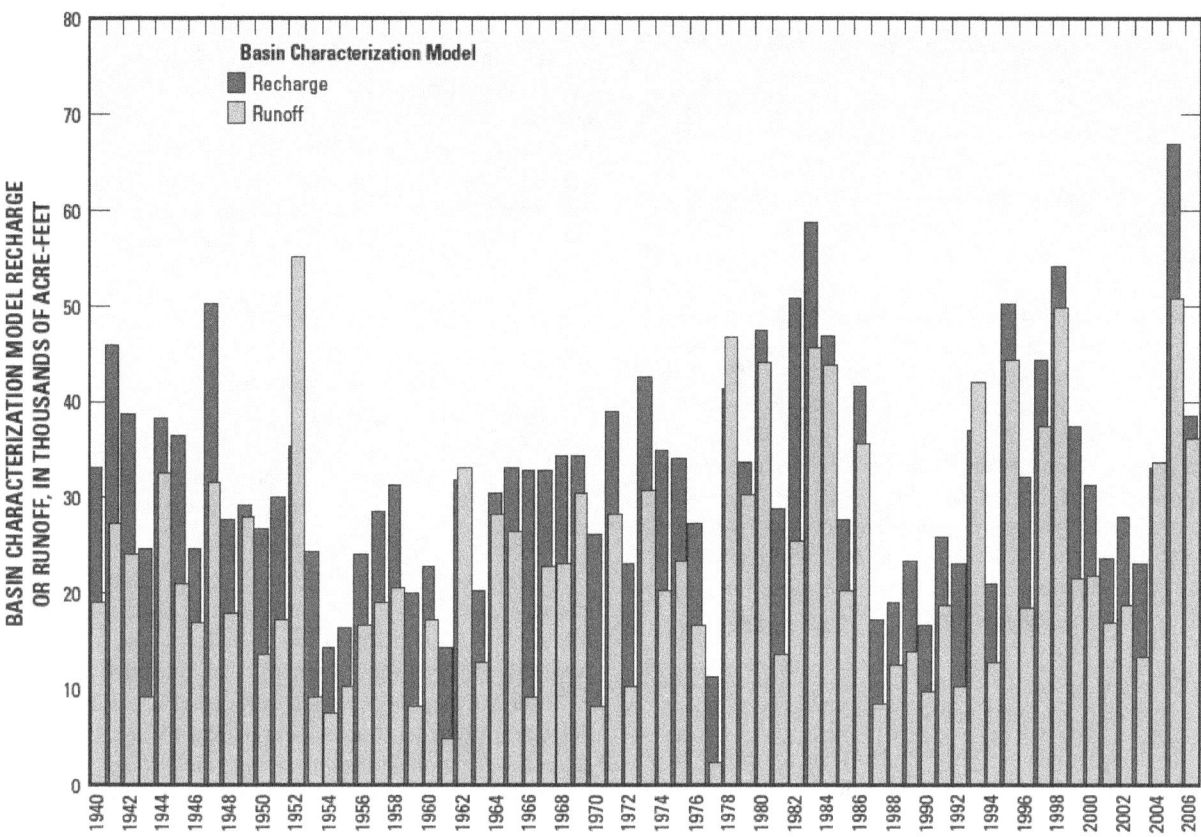

Figure 9. Annual recharge from precipitation and runoff estimated by the Basin Characterization Model for the period 1940–2006 for Rush Valley, Tooele County, Utah.

ring north and about 11,000 acre-ft/yr occurring south of the groundwater divide (table 4).

As a first step in calculating ETg, total ET was calculated for the area where groundwater is shallow enough to be transmitted by ET. This area was subdivided according to similar ET-related characteristics such as vegetation type and density, and land cover, and appropriate ET rates were applied to each zone. Average annual precipitation (1971–2000) was determined from PRISM model data (Daly and others, 2008) that was resampled to 10-m resolution. Precipitation was subtracted from the total ET to arrive at an annual estimate for ETg.

Agricultural fields exist within the ET area (area of shallow groundwater) where delivered or pumped irrigation water is sufficient to meet the consumptive requirement of the crops. Water used for irrigation in these areas is either surface water diverted from nearby mountain streams or groundwater withdrawn from wells. The contributions of both of these sources are accounted for elsewhere in the groundwater budget. Thirty percent of surface water applied for irrigation becomes recharge and is added to the groundwater budget; the remainder is consumed as ET by the crops. Groundwater from wells that is applied to crops is subtracted from the budget as the "well withdrawal" component of discharge. For these reasons, the ET estimated for these fields was omitted from

the total calculated ET. Agricultural fields classified as "sub-irrigated" were not omitted from the ETg calculation.

The outer boundary of the ET area delineated in this study approximates the extent of the phreatophytic vegetation (including areas of moist bare soil) where groundwater may be transferred to the atmosphere by ET. Results of ET studies in areas of Nevada and California (Nichols, 2000; Berger and others, 2001) suggest that most ETg occurs when the water table is within 15 to 20 ft of land surface and that phreatophytes commonly grow in areas where the depth to water is within about 40 ft of land surface (DeMeo and others, 2008). The boundary used to calculate ETg in this study was modified from one delineated for large-scale ET areas in the Great Basin (Medina, 2005) to coincide with the area where groundwater is shallow according to the water-level surface maps developed during this study. After refinement using 1-m resolution National Agricultural Imagery Program (NAIP) Digital Orthorectified Aerial Images from 2006 (U.S. Department of Agriculture, 2006) and field verification, the final ETg area boundary approximately corresponded to where groundwater is within about 25 ft of land surface.

Prior to assigning ET rates, the ET area boundary was subdivided into smaller zones (ET units) based on vegetation and land-cover characteristics determined by using existing land-cover and land-use data. An ET unit is an area of similar vegetation or land-cover characteristics that is assigned one ET rate. Southwest Regional Gap Analysis Program (SWReGAP) land-cover data (U.S. Environmental Protection Agency, 2007) and Utah water-related land-use survey (WRLU) data from 1993 (Utah Department of Natural Resources, 2004) were used to identify subareas of common vegetation or land cover. These data provided subarea boundaries that could be verified or slightly modified using the NAIP imagery and grouped into ET units. The Utah WRLU survey boundaries, considered more accurate than the SWReGAP data because they are field mapped and updated approximately once per decade, were used preferentially in developed areas. Evapotranspiration rates reported in recent literature (Nichols, 2000; Berger and others, 2001; Reiner and others, 2002; Cooper and others, 2006; and Moreo, 2007) for vegetation types, land cover, open water, and climate similar to those in the study area were assigned to 11 ET units in Rush Valley (fig. 10). Much of the area where ETg is occurring in Rush Valley is undeveloped and covered by the ET units designated as moderately dense to dense desert shrubland. The ET unit that occupies most of the developed area is pasture/range. Table 6 contains the values used in the ET calculations.

Mountain Springs and Baseflow to Mountain Streams

Groundwater that discharges from mountain springs or as baseflow to mountain streams is estimated to be about 11,000 acre-ft/yr (table 4). Discharge from significant mountain springs enters stream channels in their respective drainages and is included (as part of the baseflow component) in the gaged or estimated annual streamflow for the individual streams listed in table 1. Estimates were made as described in the Streamflow subsection of this report.

Well Withdrawals

Groundwater withdrawal by wells in Rush Valley was estimated to be between 4,200 and 4,800 acre-ft/yr in the mid-1960s (Hood and others, 1969) and is assumed to have remained relatively constant through the mid-1990s (fig. 11). Although well withdrawal fluctuates from year to year, it has increased by about 40 percent in recent years, from about 4,600 acre-ft in 1995 to a maximum of about 6,500 acre-ft in 2007 (table 7) and averaged about 6,200 ft/yr (2004–2008; table 4). Although a small amount of well withdrawal is dispersed across sparsely or unpopulated areas for irrigation or stock watering, most is used for irrigation that is concentrated around the three populated areas of Stockton, Rush Valley (Clover/St. John), and Vernon. Well withdrawal currently makes up about 15 percent of the total discharge in Rush Valley. This fraction is expected to increase in the future as some municipal supplies are converted from surface water to groundwater and as residential growth continues in Rush Valley.

Subsurface Outflow to Tooele and Cedar Valleys

The water-level surface map (fig. 7) indicates that groundwater in basin-fill deposits in Rush Valley discharges eastward from Rush to Cedar Valley in the vicinity of Five Mile Pass and northward from Rush to Tooele Valley in the vicinity of the Stockton Bar. The amount of water (Q) discharged from Rush Valley across each of these areas is calculated from the equation (Freeze and Cherry, 1979):

$$Q = T \left(\frac{dh}{dl}\right) L \qquad (2)$$

where:

T is the transmissivity (in ft²/day, estimated from aquifer tests),

$\dfrac{dh}{dl}$ is the hydraulic gradient or slope of the water-level surface (dimensionless), and

L is the length of the area across which the discharge is occurring (in ft).

The amounts of subsurface discharge calculated using equation 2 are 310 acre-ft/yr from Rush to Cedar Valley and 620 acre-ft/yr from Rush to Tooele Valley. These calculations were based on hydraulic gradients determined from the water-level surface map (fig. 7) using flow nets (Domenico and Schwartz, 1998) constructed between the 4,900- and 5,000-ft water-level contours in both areas and the median T value (290 ft²/d) for basin-fill deposits (table 3). These estimates fall

Table 6. Evapotranspiration unit rates and areas used to calculate average annual evapotranspiration of groundwater in Rush Valley, Tooele County, Utah.

[ET, evapotranspiration]

ET unit	ET rate, area weighted average, in feet	Range of ET rate, in feet	Acres	Total ET, in acre-feet
Dense desert shrubland[1]	1.2	1.0–1.8	16,174	20,056
Moderately dense desert Shrubland[1]	1.1	0.7–1.5	13,802	14,768
Sparse desert shrubland[1]	0.9	0.5–1.1	1,100	990
Pasture/range[2]	2.0	0.8–3.1	15,696	30,895
Grassland[1]	2.1	1.6–2.7	1,258	2,693
Marshland[1]	4.1	3.6–4.6	2,567	10,448
Meadowland[1]	2.6	2.2–3.3	256	664
Grain[2]	1.7	0.5–3.0	228	396
Moist bare soil[1]	2.0	1.7–2.3	139	279
Alfalfa[2]	2.5	1.1–3.9	112	278
Open water[1]	5.1	4.6–5.6	31	159

Sum of total ET   81,625
– Precipitation over ET area   56,325

= ET from groundwater (rounded)   25,000

[1]ET rates and ranges for these ET units are summarized in Welch and others (2007), p 54–56

[2]ET rates and ranges for these ET units from Utah State University (1994), table 25 Ranges (high and low) are the average ET rate ± 1-σ

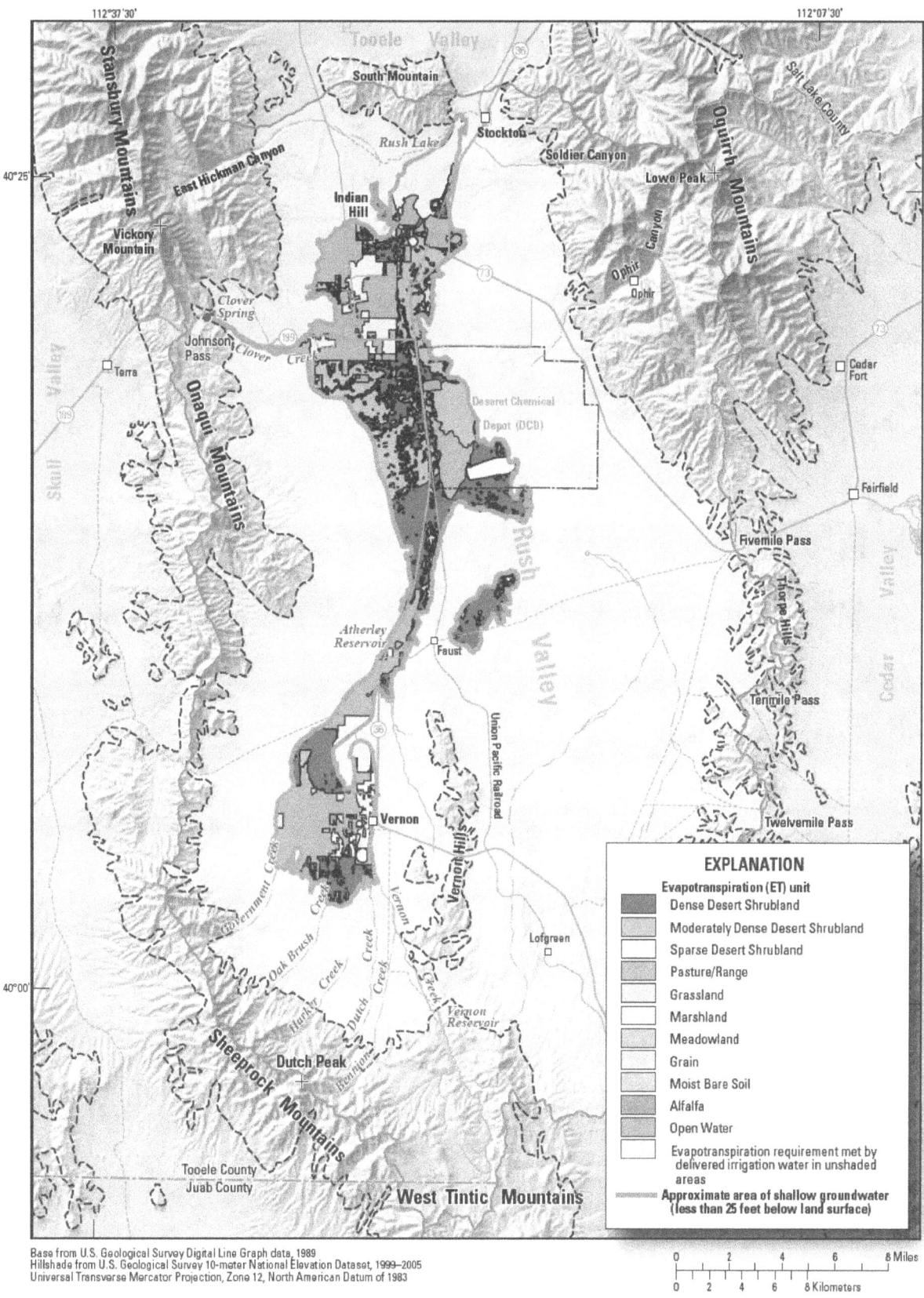

Figure 10.   Locations and classification of evapotranspiration units used to calculate average annual evapotranspiration of groundwater in Rush Valley, Tooele County, Utah.

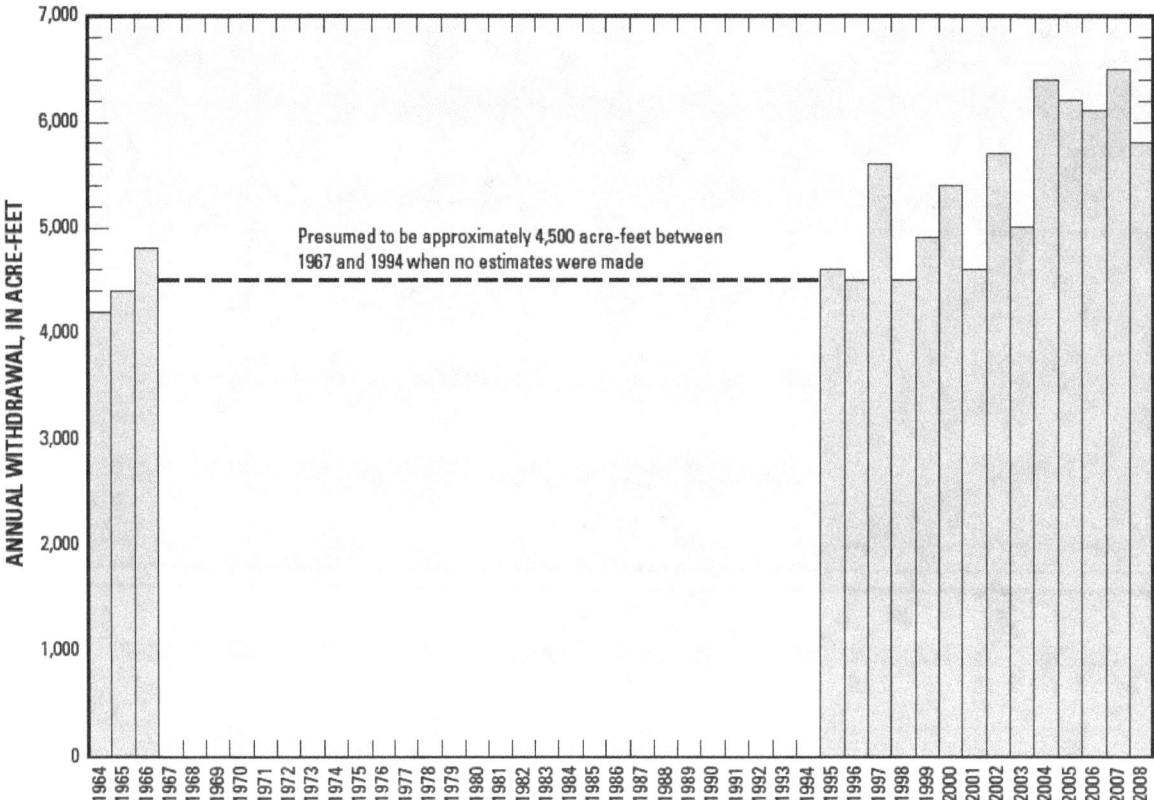

Figure 11.   Annual withdrawal of groundwater by wells, 1964–2008, Rush Valley, Tooele County, Utah.

within the range of previous estimates of discharge from Rush to Cedar Valley and from Rush to Tooele Valley, which range from "small" to about 5,000 acre-ft/yr (Thomas, 1946; Gates, 1965; Hood and others, 1969, Razum and Steiger, 1981; and Stolp and Brooks, 2009) The estimate of subsurface discharge from Rush to Cedar Valley is in agreement with that made by Jordan and Sabah as part of an ongoing study (L. Jordan, written commun., 2010). The estimate of subsurface discharge from Rush to Tooele Valley is in agreement with the "small but significant" amount postulated by Hood and others (1969) but less than that used in the later modeling study by Razum and Steiger, 1981.

Water-Level Fluctuations

Water levels in wells fluctuate in response to imbalances between groundwater recharge and discharge. Water levels rise when recharge exceeds discharge for a period of time and decline when the opposite occurs. Variations in recharge and discharge are driven by natural and anthropogenic (human-induced) processes. Examples of natural processes are recharge from the infiltration of precipitation and evapotranspiration of groundwater in a marsh or wetland. The infiltration of unconsumed water applied to irrigate crops or groundwater withdrawal by wells are examples of anthropogenic processes. Long-term and seasonal water-level

Table 7.   Annual withdrawal of groundwater used for irrigation, industry, public supply, and domestic and stock use by wells, 1964–2008, in Rush Valley, Tooele County, Utah.

[All withdrawal values are in acre-feet Values for 1964–1966 from Hood and others (1969) Values for 1995–2008 from Burden and others (2009) —, no data]

Year	Irrigation	Industrial use	Public supply	Domestic and stock	Total well withdrawal
1964	—	—	—	—	4,200
1965	—	—	—	—	4,400
1966	—	—	—	—	4,800
1995	—	—	—	—	4,600
1996	3,900	140	440	30	4,500
1997	4,300	1,000	280	30	5,600
1998	3,900	200	330	30	4,500
1999	4,600	0	300	30	4,900
2000	4,800	280	290	30	5,400
2001	4,100	170	250	30	4,600
2002	5,200	180	270	30	5,700
2003	4,500	170	300	30	5,000
2004	5,900	170	280	30	6,400
2005	5,700	170	290	30	6,200
2006	5,600	170	270	30	6,100
2007	5,900	250	340	30	6,500
2008	5,200	250	290	30	5,800

changes were examined for selected wells within and adjacent to the Rush Valley study area (fig. 12). Long-term water-level fluctuations are presented for 12 wells where repeated measurements have been made for various periods of time to illustrate the groundwater system's response to interannual variations in recharge and discharge (fig. 13). Long-term water-level data were filtered to include only March water-level measurements for each year. For years when a March measurement was not available, the closest springtime measurement was used. Monthly water levels are presented for 11 wells to illustrate the groundwater system's response to seasonal recharge and discharge (fig. 14). All water-level data are available through the USGS National Water Information System (NWIS) database (http://waterdata.usgs.gov/nwis).

All wells with long-term water-level data in northern Rush Valley are completed in the shallow (less than 200 ft) UBFAU (fig. 13).Water levels in wells (C–4–5)33cca–1, (C–5–5)2bcb–1, (C–5–5)5bdb–1, (C–5–5)20daa–2, and (C–5–5)31dbd–1 and –2 rose and fell within a couple of years following periods of above average precipitation (1982–1983 and/or 1996–1998) (fig. 13), indicating that this part of the groundwater system responds to recharge on relatively short timescales.

Long-term water-level data in the Vernon area are available for wells completed in several hydrogeologic units. The UBFAU wells C–8–5)20cdd–1, (C–8–5)31ccd–5, and (C–9–5)6aab–1 showed 5- to 15-ft water-level rises following the multiyear period of above average precipitation in the early to mid-1980s (fig. 13). Water-level response to the early 1980s wet period in these wells was more delayed than in wells completed in the UBFAU in northern Rush Valley, possibly due to the greater distance of these wells from the adjacent mountain recharge area. Well (C–8–5)7ddd–2 is finished in permeable gravels of the UBFAU directly overlying fractured bedrock of the UCAU. The water level in this well is 60 to 110 ft below the water levels in nearby UBFAU wells, suggesting that this well represents conditions in the UCAU aquifer (fig. 12, table A1–2). Water levels in this well rose more than 24 ft from 1983 to 1990, possibly indicating a lagged response to the early to mid-1980s wet period. Since that time, the water level in this well has declined more than 40 ft while showing little to no recovery during subsequent periods of above average precipitation (1996–98 and 2004–05). This water-level decline may be the result of large withdrawals for irrigation that occur north of Vernon from several high-volume wells pumping from fracture zones in the UCAU. Well (C–7–5)32bdd–1 is finished in the low-permeability semiconsolidated LBFAU and exhibits minimal long-term water-level fluctuations.

Long-term water levels near the eastern border of the study area show the least variation. Well (C–7–3)30aac–1 is completed in the UBFAU. This well exhibits generally small water-level changes from year to year with a maximum water-level change of about 7 ft occurring during 1984–85. However, nearly all of the water-level variation in this well is attributable to drawdown from pumping each spring. The

timing of stock well pumpage in southeastern Rush Valley is less predictable than that of irrigation wells in Rush Valley. In some years, a particular stock well may not even be used. Because well (C–7–3)30aac–1 is sometimes pumped prior to March water-level measurements being made, the measurements likely represent water levels that are at various stages of recovery from year to year. Water-level fluctuations in well (C–7–2)29dbc–1, located about 4 mi east of the study area, were included to examine conditions along a flow path where groundwater from Rush Valley likely moves into Cedar Valley. Annual water-level fluctuations in this well are small, generally less than 0.5 ft, indicating that a nearby area of substantial recharge probably does not exist. Water levels in this well showed a rise of about 3.5 ft over a 10-year period from about 1985 to 1995 that may be a long-term damped response to the mid-1980s period of above average precipitation.

Water-level measurements were made on a monthly basis during 2008–09 to monitor the groundwater system's response to seasonal recharge and discharge (fig. 14). Measurements were not made when a well was pumping at the time of a monthly visit.

In northern Rush Valley, wells (C–4–5)29bdc–2 (south of South Mountain) and (C–4–5)13bad–1 (north of the Stockton Bar) are located near adjacent bedrock mountain highlands (fig. 12). Both of these wells exhibit little to no seasonal water-level fluctuation, indicating that they receive very little (if any) nearby recharge and that they are not affected by nearby pumping wells. The abrupt one-time rise in water level in well (C–4–5)27cdb–1 coincides with the end of use of this domestic well and likely represents a return to local conditions unaffected by pumping. Well (C–5–5)32cac–2, located in Clover, illustrates seasonal drawdown associated with summertime pumping and subsequent recovery during the winter months.

Monthly water levels in several Vernon area wells appear to respond to seasonal pumping. Wells (C–9–5)5bbc–1 and (C–9–5)6aab–1 are located in an agricultural area south of Vernon that is irrigated with groundwater pumped from the UBFAU (fig. 12). Both of these wells show a clear decline and recovery pattern coinciding with the 2008 and 2009 irrigation seasons (fig. 14). Well (C–7–5)32bdd–1 is finished in the LBFAU and experiences similar but less pronounced water-level fluctuations. Well (C–8–5)20ccd–1 is completed in the UBFAU. Although its seasonal water-level record is incomplete and affected by intermittent pumping (as indicated by missing data), the small water-level variation does not indicate interference from nearby irrigation-related pumping that occurs in both the UBFAU and the UCAU.

Wells (C–6–4)35bac–1 and (C–7–3)30aac–1 in southeastern Rush Valley and (C–7–2)29dbc–1 in Cedar Valley (fig. 12) display little to no seasonal water-level fluctuation (fig. 14). These wells are located far from any recharge source, and groundwater pumping is minimal in this part of Rush Valley. Well (C–6–4)35bac–1 is located on the flank of the southern Oquirrh Mountains near the toe of

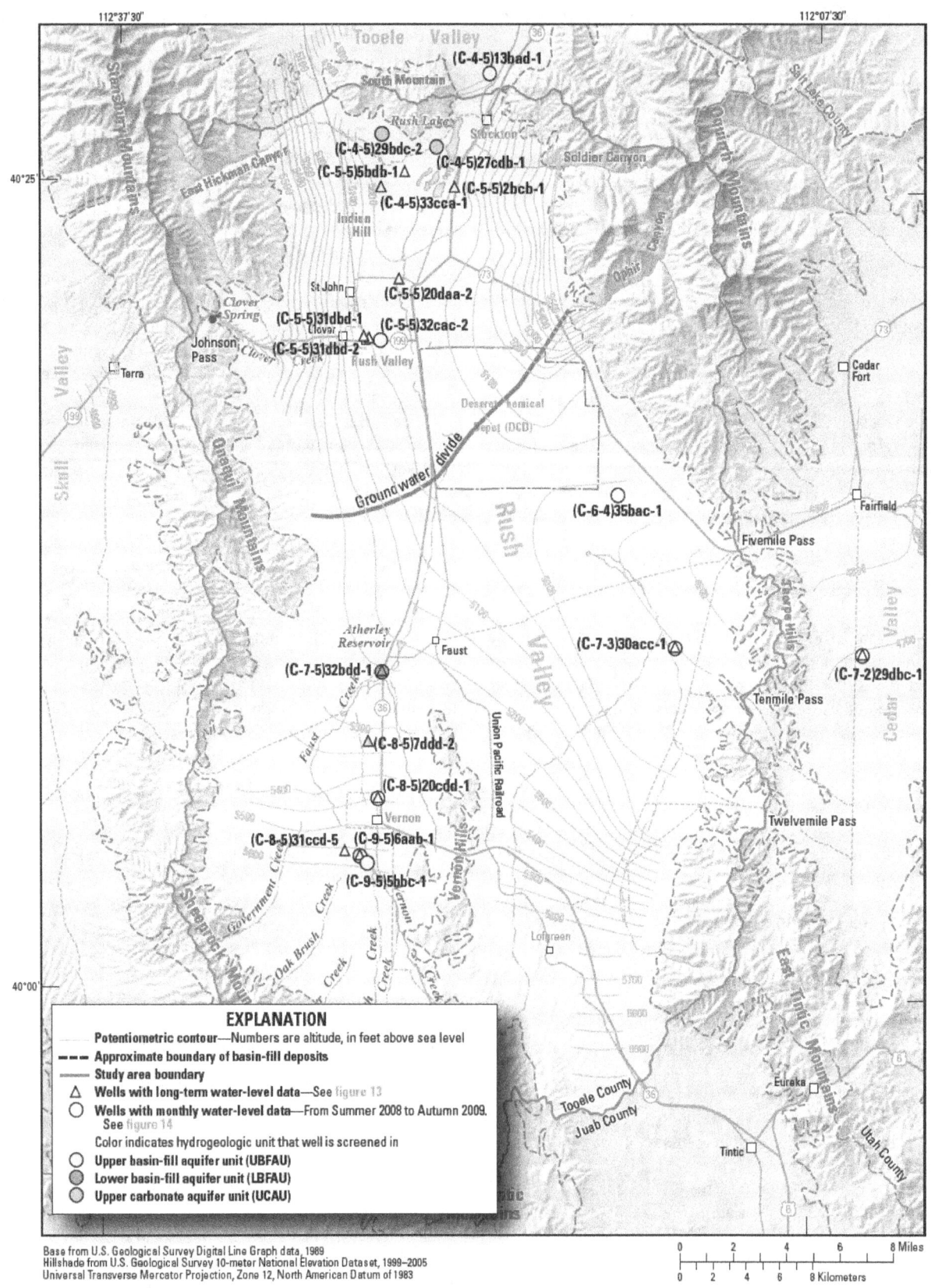

Figure 12.   Locations of selected wells with long-term and monthly water-level data, Rush Valley, Tooele County, Utah.

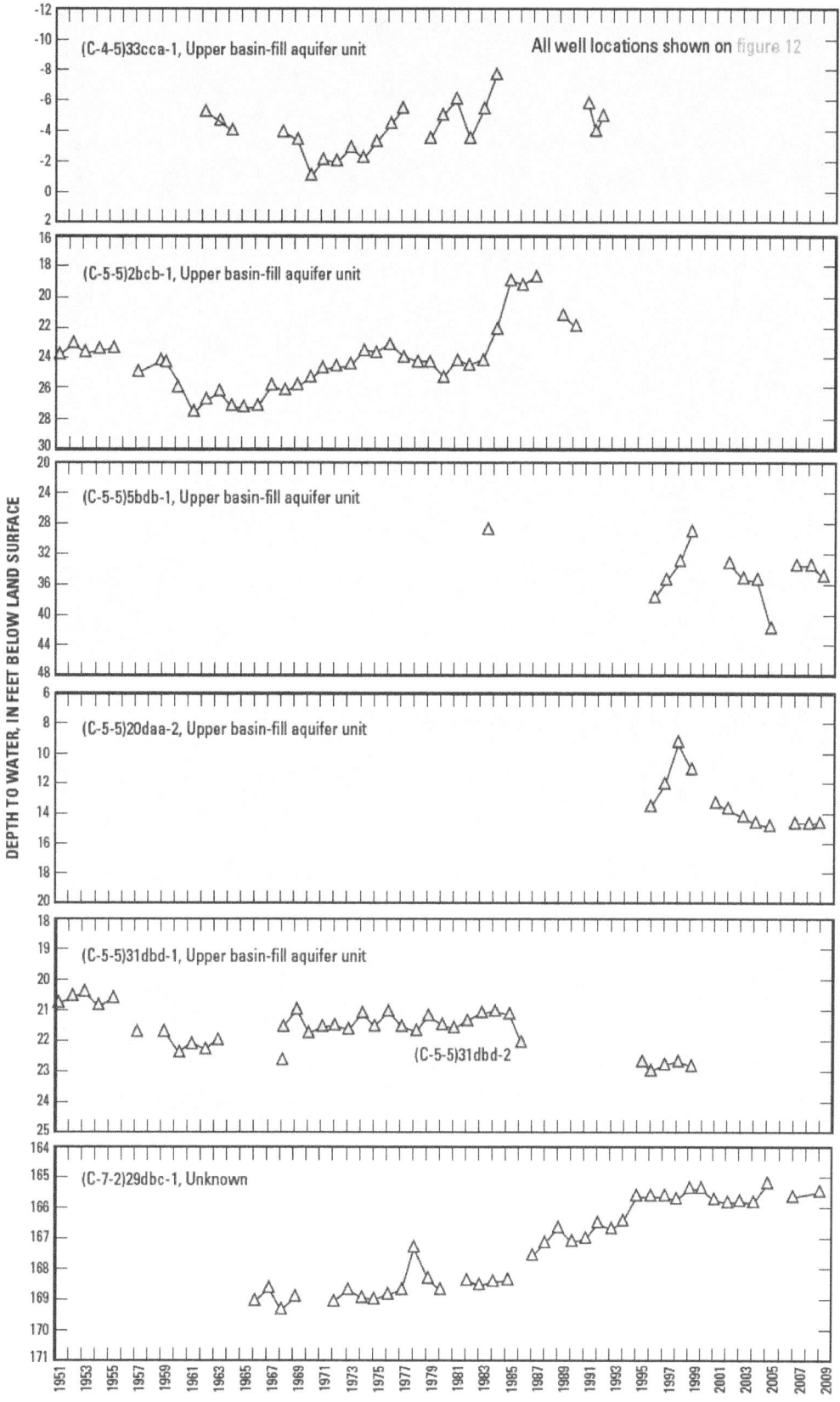

Figure 13.   Long-term water-level fluctuations in selected wells, Rush Valley, Tooele County, Utah.

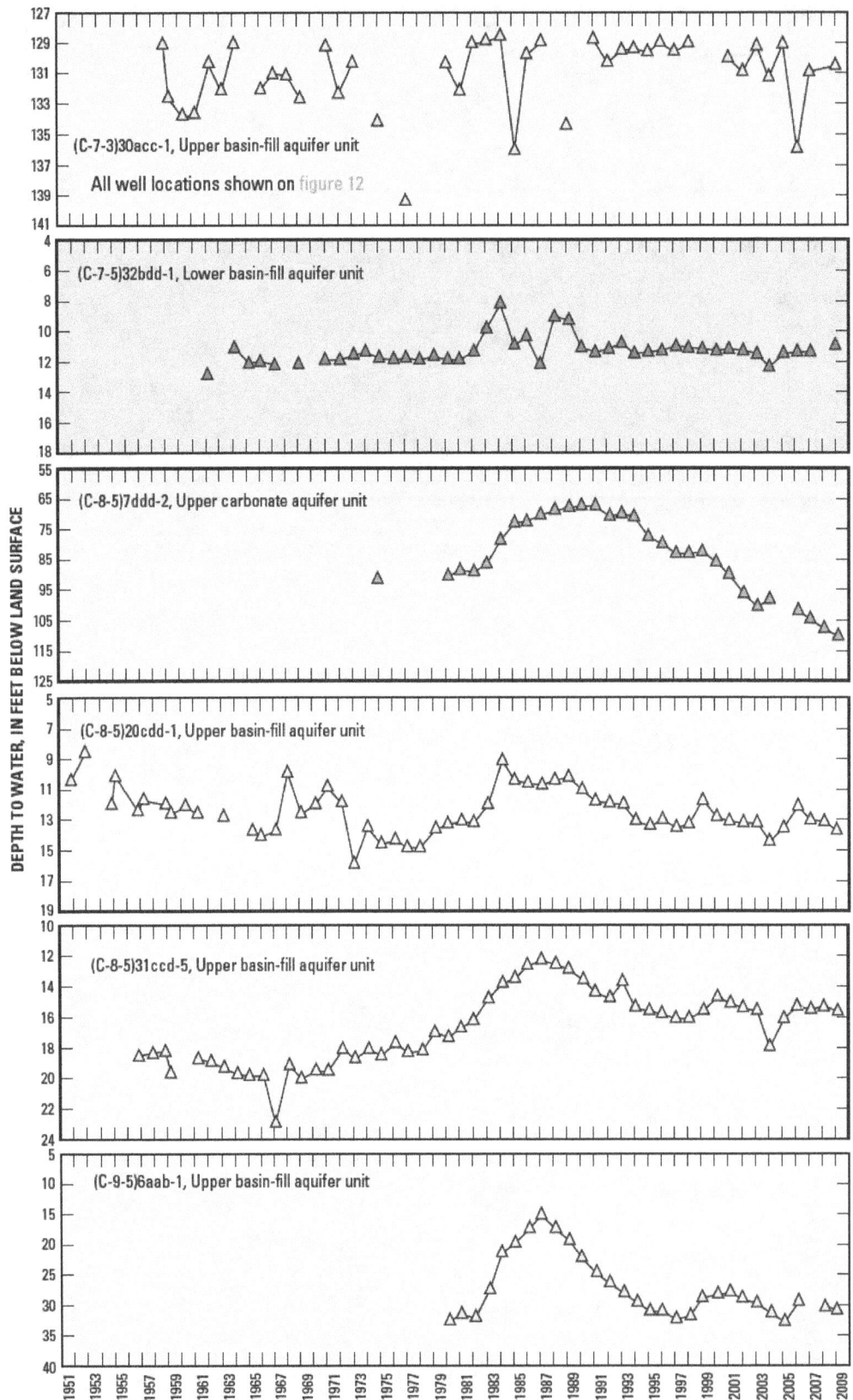

Figure 13.   Long-term water-level fluctuations in selected wells, Rush Valley, Tooele County, Utah.—Continued

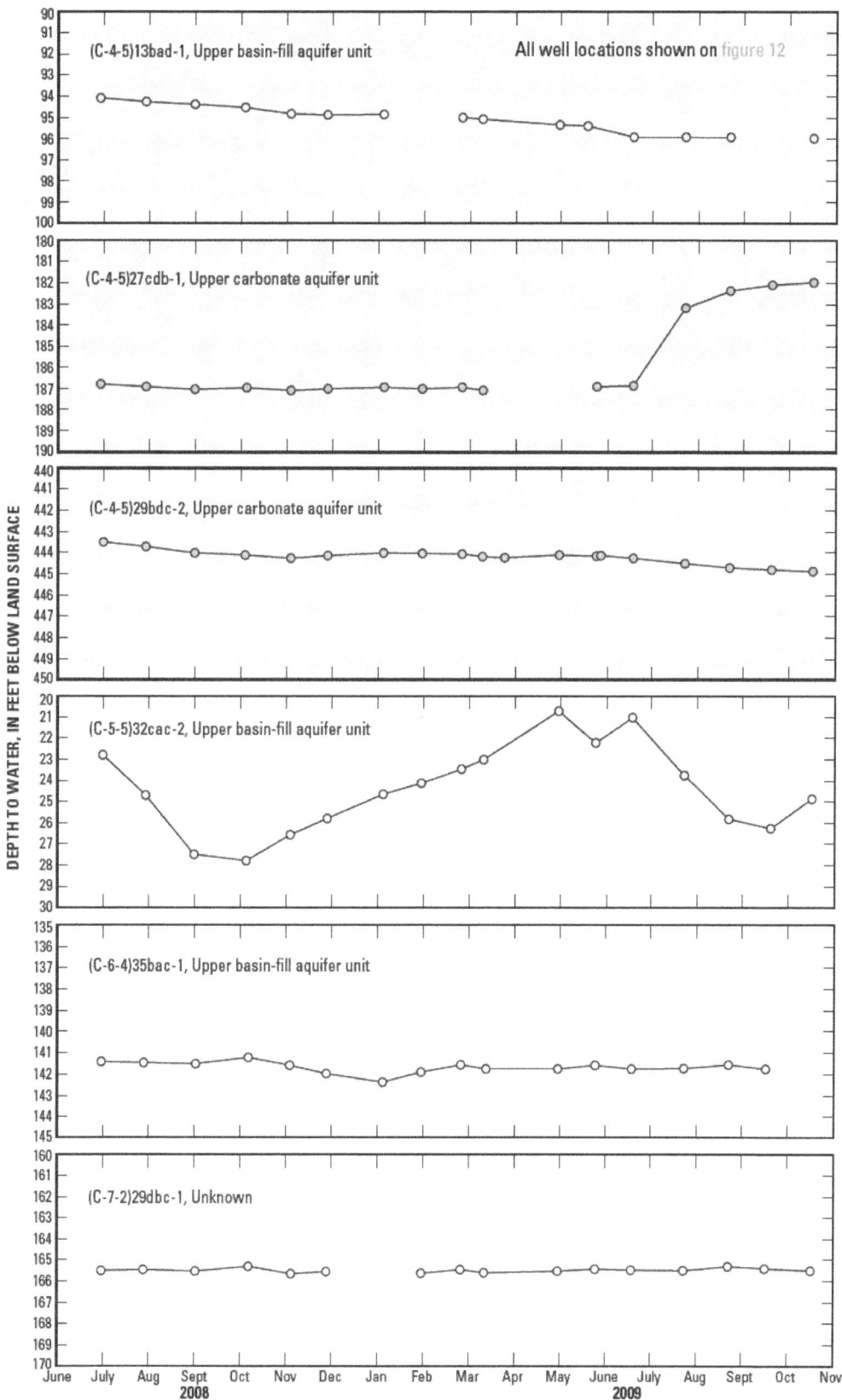

Figure 14. Monthly water-level fluctuations in selected wells during 2008–09 within and adjacent to Rush Valley, Tooele County, Utah.

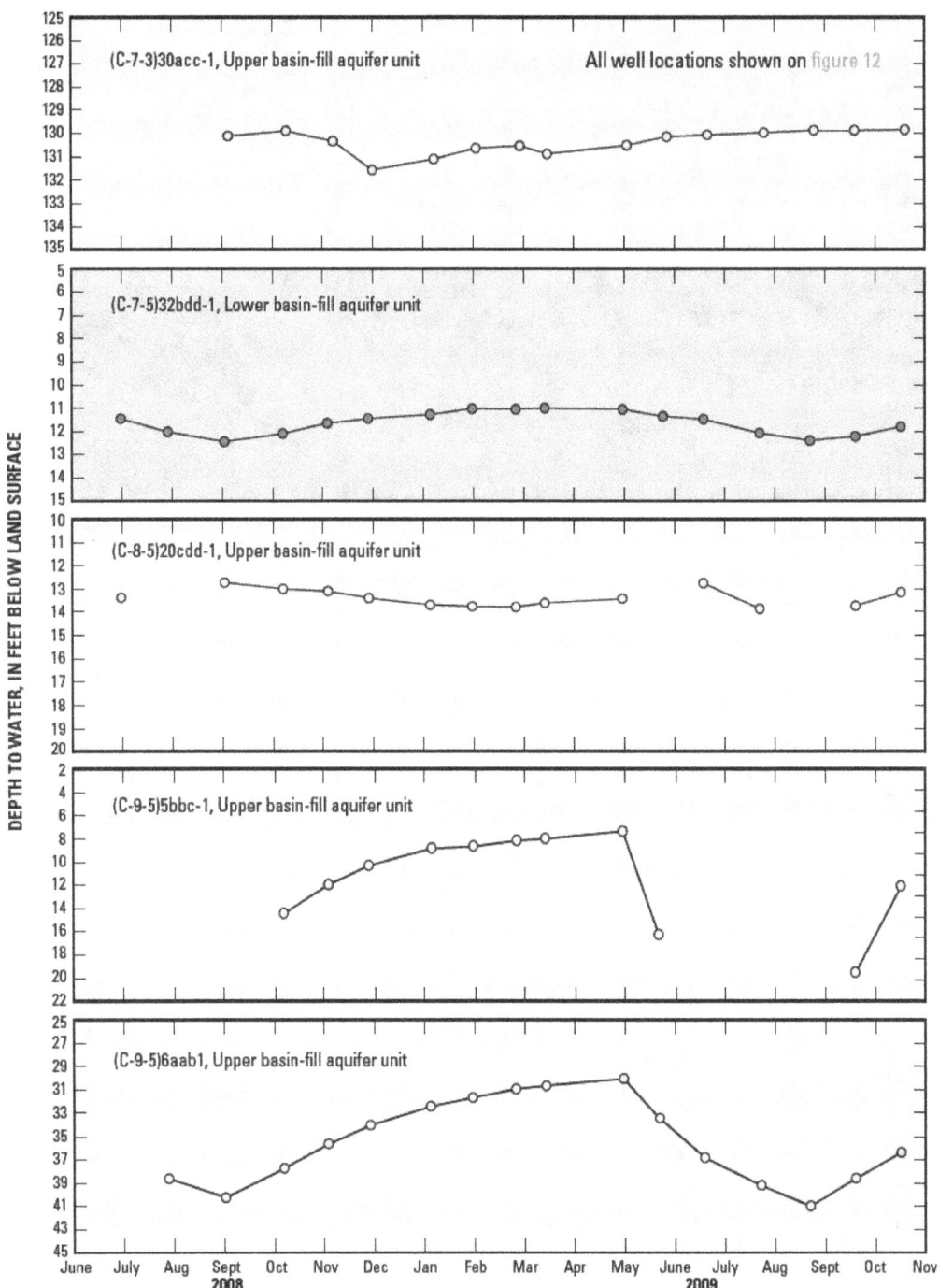

Figure 14.   Monthly water-level fluctuations in selected wells during 2008–09 within and adjacent to Rush Valley, Tooele County, Utah.—Continued

an alluvial fan. The lack of any significant water-level rise in this well during 2008–09 may indicate that geologic structure prevents recharge in the southern Oquirrh Mountains from moving westward into Rush Valley. The slight drawdown and recovery pattern in stock well (C–7–3)30aac–1 is likely a response to pumping of this well, which occurs less regularly than the irrigation-related pumping in other parts of Rush Valley.

Groundwater Geochemistry

Water samples were collected from 25 sites in Rush Valley and included observation, municipal, domestic, irrigation, and stock wells and one perennial spring (tables 8 and A1–2). The water samples were analyzed for major ions, nutrients, and selected trace metals to characterize general geochemistry and patterns of water quality. Water samples also were analyzed for a suite of environmental tracers that included the stable isotopes of oxygen, hydrogen, and carbon, dissolved noble gases, and radioactive isotopes of carbon (^{14}C) and hydrogen. (tritium, ^{3}H). These environmental tracers were used to investigate sources of recharge, groundwater-flow paths, ages, and traveltimes and to support the development of a conceptual model of the basin-wide groundwater system. All groundwater geochemical data are available through the USGS NWIS database (http://waterdata.usgs.gov/nwis).

Sample Collection and Analysis

Samples were collected from wells using either the well's dedicated pump or a portable submersible pump. At the spring and one flowing well, water samples were collected under natural free-flowing conditions. Water samples from wells were collected after being purged of three casing volumes of water. Sample water was collected from an outlet as close to the wellhead as possible and before entering any storage or pressure tanks.

Field parameters measured during water-sample collection include specific conductance, pH, temperature, dissolved oxygen, and total dissolved-gas pressure. These parameters were measured using a calibrated multimeter probe following USGS protocols (Wilde and Radtke, 1998). Samples for dissolved major ions and nutrients were filtered with a 0.45-μm filter. The cation subsample was preserved with nitric acid. Dissolved major-ion and nutrient analyses were done by the USGS National Water Quality Laboratory in Denver, Colorado. Unfiltered samples for stable isotopes of oxygen and hydrogen were collected in 60-mL glass containers, sealed with polyseal caps leaving no air space, and analyzed by the USGS Stable Isotope Laboratory in Reston, Virginia. Unfiltered samples for tritium were collected in 500-mL or 1-L polyethylene bottles, sealed with no air space in the container, and analyzed by the University of Utah Dissolved Gas Service Center in Salt Lake City. Samples for carbon-14 (^{14}C) and

stable isotopes of carbon were filtered (0.45 μm) and collected in 500-mL or 1-L glass bottles. The bottles were filled from the bottom and allowed to overflow for several volumes in order to rinse the bottles while minimizing contact with the air, sealed with polyseal caps, and analyzed by the National Ocean Sciences Accelerator Mass Spectrometry Facility (NOSAMS) at the Woods Hole Oceanographic Institution in Woods Hole, Massachusetts.

Dissolved-gas samples were collected either as water samples sealed in copper tubes as described by Stute and Schlosser (2000) or as gas samples collected with diffusion samplers as described by Sheldon (2002). The copper tube method consists of attaching a 30-in.-long section of 3/8-in.-diameter copper tubing to a sampling port at the wellhead, allowing the tube to flush with well water until all air bubbles have been evacuated, then sealing both ends with clamps. The diffusion sampler method was used at wells and springs where either in-situ placement or uninterrupted flow using a flow-through chamber was possible for a minimum period of 24 hours. The diffusion sampler is constructed of 1/8-in.-diameter copper tubing and a semipermeable gas diffusion membrane. The flow-through chamber is an airtight chamber connected to a discharge point at the wellhead, allowing water to flow through the chamber and past the membrane. After 24 hours, when the gases in the diffusion sampler had equilibrated with the dissolved gases in the sample water, the sampler was removed from the well or spring and immediately sealed. Dissolved-gas concentrations were analyzed by the University of Utah Dissolved Gas Service Center using both quadrupole and sector-field mass spectrometers. The analysis provides the relative mole fractions of gases dissolved in a sample and the dissolved-gas concentrations are calculated by using Henry's Law relations and field measurements of total dissolved-gas pressure and water temperature.

Environmental Tracer Methods

Environmental tracers used in this study assist in developing and refining conceptual models of groundwater-flow systems. They can be used to investigate sources of recharge and to determine rates of movement and ages of groundwater. They also help to refine groundwater-flow paths originally delineated using water-level surface maps.

Tritium and Helium

Tritium and helium isotopes were used in this study to examine the age of groundwater samples. Tritium (^{3}H) is a radioactive isotope of hydrogen with a half-life of 12.32 years that decays to tritiogenic helium-3 ($^{3}He_{trit}$). Tritium is present in water as part of the water molecule, whereas its decay product, $^{3}He_{trit}$, exists as a noble gas dissolved in water. Measured concentrations of ^{3}H and $^{3}He_{trit}$ can be used to determine the apparent age of groundwater that is less than about 60 years old. During the 1950s and 1960s, large amounts of ^{3}H were released into the atmosphere

Table 8. Measured field parameters and dissolved concentrations of major ions, nutrients, and selected metals for groundwater sampled during 2008–2010 in and around Rush Valley, Tooele County, Utah.

[Sample ID: see figure 15 for the location of sites sampled as part of this study and table A1–2 for physical information about the sampled well or spring. C, degrees Celsius; μS/cm, microsiemens per centimeter at 25 degrees Celsius; mg/L, milligrams per liter; μg/L, micrograms per liter; <, less than; E, estimated; —, not analyzed; values in red type exceed the Environmental Protection Agency maximum contaminant level or secondary standard]

Sample ID	Sample date	Air temperature, °C	Water temperature, °C	Specific conductance, μS/cm	pH, standard units	Dissolved oxygen, mg/L	Alkalinity, mg/L as CaCO₃	Bicarbonate, mg/L as HCO₃	Bromide, mg/L as Br	Calcium, mg/L as Ca	Chloride, mg/L as Cl	Fluoride, mg/L as F	Iron, μg/L as Fe	Magnesium, mg/L as Mg
1	6/1/2009	28.0	14.0	1,050	7.6	8.0	132	161	0.14	86.7	231	0.14	5	45.6
2	6/3/2009	31.0	12.6	770	7.5	8.5	138	168	0.11	82.8	134	E0.1	13	19.2
3	6/3/2009	27.5	13.0	613	7.5	9.1	142	173	0.07	68	88.8	0.12	15	16.1
4	7/29/2008	—	16.1	575	7.9	0.3	158	192	0.05	41.8	68.2	0.42	14	25.3
5	6/5/2009	24.0	10.6	1,490	7.9	0.0	74	90	0.07	109	78.3	0.91	39	66
6	5/29/2009	25.7	13.9	920	7.8	<0.1	179	218	0.11	25.5	155	1.41	67	31.1
7	4/28/2009	15.0	16.9	2,610	6.9	—	208	253	0.56	166	587	0.13	E8	68.2
8	5/6/2009	24.0	13.6	2,430	7.3	0.7	165	201	0.55	121	630	0.82	41	138
8¹	5/6/2009	—	—	—	—	—	164	200	0.56	117	630	0.84	39 3	138
9	5/15/2009	17.5	13.8	604	7.3	4.3	230	281	0.04	42	22 9	0.70	6	25.1
10	4/28/2009	18.0	17.7	902	7.8	0.0	237	288	0.08	40.8	86.8	1.22	488	23.8
11	4/22/2009	26.0	12.0	605	7.9	4.9	154	188	0.15	29.9	60.6	0.67	<4	31.5
12	4/22/2009	26.0	23.0	857	7.5	0.1	253	309	0.11	43.5	38.4	2.83	2,260	27.6
13	5/15/2009	21.0	13.0	596	7.8	0.0	218	266	0.03	28.4	31 9	0.99	77	30.9
14	5/12/2009	20.5	7.6	360	7.4	8.4	185	226	<0.02	64.6	4.85	0.15	<4	5.76
14	9/25/2009	27.0	7.5	329	7.6	8.9	161	197	<0.02	49.6	5.46	0.12	<4	10.1
15	6/2/2009	23.0	12.6	1,540	7.3	4.9	165	201	0.22	130	313	0.17	20	37.5
16	6/2/2009	12.0	13.8	972	7.4	1.6	127	155	0.18	85.5	211	0.19	152	34.7
17	5/26/2009	23.0	10.0	1,080	7.1	—	316	386	0.15	107	139	0.28	96	26.6
18	5/6/2009	21.0	21.6	1,150	7.5	1.7	144	176	0.18	52.2	206	0.52	14	28.8
19	7/29/2008	—	21.9	823	7.4	5.1	156	190	0.11	62.5	134	0.26	20	30.5
20	6/5/2009	22.0	11.4	675	7.3	0.2	255	311	E0.02	79.4	25.1	0.21	45	32
21	7/29/2008	—	11.5	549	7.3	—	196	239	0.04	53.7	41.4	0.18	<8	27.2
22	5/29/2009	32.5	14.6	535	7.6	2.7	161	196	0.04	48.5	40 2	0.51	13	26.2
23	5/12/2009	23.0	12.4	1,300	7.3	0.0	218	266	0.21	98.3	248	0.79	453	36.5
24	5/26/2009	19.3	14.3	873	7.4	4.1	165	201	0.12	74.9	148	0.42	10	22.1
25	6/2/2009	23.0	22.1	2,080	7.2	0.6	174	212	0.33	94.7	481	0.28	11	62.5
Environmental Protection Agency maximum contaminant level (MCL)														
Environmental Protection Agency secondary standard											250	2	300	

¹Replicate sample included to ensure quality control of analyses and repeatability of interpreted values
²For nitrate only

and introduced into the hydrologic cycle by above ground thermonuclear weapons testing. As a result, ³H concentrations in precipitation in the northern hemisphere peaked during 1963–64 at three orders of magnitude above natural concentrations (Michel, 1989). Comparison of reconstructed initial ³H concentrations with atmospheric concentration data is a tool that can distinguish between groundwater recharged before or after the beginning of weapons testing in the mid-1950s. By using the concentrations of both ³H and its decay product, ³He$_{trit}$, the age of groundwater (time elapsed since recharge) can be refined to an apparent recharge year. These ages are referred to as "apparent" because they can differ from the true mean age of the sample if it contains a mixture

of water of different ages. Mixtures of modern (post- mid-1950s recharge) and premodern (pre- mid-1950s recharge) water typically have apparent ³H/³He ages that represent the age of the young fraction of the sample because dilution with premodern water will leave the ratio of ³H to ³He$_{trit}$ virtually unchanged. Details of this groundwater-dating method are presented in Solomon and Cook (2000).

Tritium concentrations typically are reported in tritium units (TU), where one TU is equivalent to one molecule of tritiated water (³H¹HO) in 10¹⁸ molecules of ¹H₂O. In a sample of premodern groundwater, ³H will have decayed from background "prebomb" concentrations of about 6– 8 TU to less than 0.3 TU, which is approaching the analytical

Table 8. Measured field parameters and dissolved concentrations of major ions, nutrients, and selected metals for groundwater sampled during 2008–2010 in and around Rush Valley, Tooele County, Utah.—Continued

[Sample ID: see figure 15 for the location of sites sampled as part of this study and table A1–2 for physical information about the sampled well or spring. C, degrees Celsius; µS/cm, microsiemens per centimeter at 25 degrees Celsius; mg/L, milligrams per liter; µg/L, micrograms per liter; <, less than; E, estimated; —, not analyzed; values in red type exceed the Environmental Protection Agency maximum contaminant level or secondary standard]

Manganese, µg/L as Mn	Potassium, mg/L as K	Silica, mg/L as SiO$_2$	Sodium, mg/L as Na	Solids, residue on evaporation at 180°C, mg/L	Sulfate, mg/L as SO$_4$	Ammonia, mg/L as N	Nitrate plus nitrite, mg/L as N	Nitrite, mg/L as N	Ortho-phosphate, mg/L as P	Total nitrogen, mg/L as N	Arsenic, µg/L as As	Molybdenum, µg/L as Mo	Selenium, µg/L as Se	Uranium, µg/L as U
0.9	1.8\	16	35.7	671	25.2	<0.02	3.13	<0.002	0.013	3.17	1.4	0.4	0.88	0.91
E0.2	1.49	17.5	34.3	531	33	<0.02	1.53	<0.002	0.017	1.52	2.3	0.3	0.83	1.16
0.3	1.38	16.7	29.6	399	24.8	<0.02	1.42	<0.002	0.016	1.42	1.7	0.5	0.66	1.1
1.2	2.23	16.1	31	321	24.3	—	0.16	—	0.014	—	9.5	2.3	0.23	2.22
56.9	35.8	42.1	99.4	1,140	626	0.698	<0.04	E0.001	0.022	0.71	13.8	5.1	0.37	0.64
3.4	5.86	63.3	109	564	45.9	0.036	<0.04	<0.002	0.037	<0.1	35.6	10.2	<0.06	0.21
<0.40	10.1	62.3	231	1,580	173	<0.02	0.76	<0.002	0.038	0.79	9.9	4.7	1.9	8.56
2.2	11.3	68.8	105	1,590	64.2	E0.012	0.87	E0.001	0.032	0.92	14.9	1.2	2.7	4.71
2.3	11	66.5	106	1,400	65.7	E0.013	0.87	E0.001	0.031	0.89	15	1.3	2.8	4.8
0.3	6.51	66.2	44.8	416	44.3	<0.02	1.94	<0.002	0.038	1.95	26.3	8	3.8	4.87
23.2	11.9	62	103	516	77.8	0 139	<0.04	<0.002	0.03	0.16	2.1	1.6	<0.06	0.14
<0.2	5.25	50.3	40.1	375	32.1	<0.02	5.97	<0.002	0.036	5.92	27.2	3.6	2.5	4.65
45.4	2.61	11.8	95.5	508	121	0 578	<0.04	<0.002	E0.007	0.56	6.0	7.7	<0.06	5.71
124	5.12	53.5	54.6	385	60.4	0.045	<0.04	<0.002	0.06	<0.1	21.3	6.4	0.06	4.53
<0.2	0.42	7.01	4.22	213	4.1	<0.02	0.19	<0.002	0.01	0.25	0.6	0.3	0 23	0.69
<0.2	0.46	6.22	5.55	181	5.16	—	0.23	—	E0.007	—	0.5	0.2	0.49	0.73
6.1	2.32	19.8	88.1	1,100	40.6	<0.02	8.65	0.002	0.12	9.03	0.7	0.3	2	1.61
59.6	2.74	19	40.2	675	23.3	<0.02	1.59	0.002	0.01	1.6	0.4	0.8	1.4	1.29
24.2	1.31	18.5	77.1	639	50.7	<0.02	1.37	<0.002	0.017	1.43	2.1	0.9	2	3.92
0.9	11.9	19.2	120	637	98.6	E0.010	0.06	<0.002	0.011	E0.05	3.5	2	0.66	1.04
0.8	3.16	21.8	44.4	453	36.5	—	0.49	—	0.02	—	1.2	1	0.57	1.43
2.3	1.03	13.2	16.7	426	73	<0.02	0.18	0.003	0.008	0.17	2.4	4.6	28.7	5.71
<0.4	1.18	10.7	14	303	21.7	—	1.32	—	0.013	—	1.5	0.8	1 5	1.82
0.4	1.53	16.4	20.1	317	49.4	<0.02	0.45	<0.002	0.011	0.45	0.8	2.4	0.97	3.07
286	6.44	31.3	98.9	803	56.2	0.918	<0.04	E0.001	0.036	0.99	10.2	1.8	0.25	0.75
93.4	6.39	40.3	60.2	560	36.3	<0.02	2.99	<0.002	0.021	3.35	1.6	1.8	5.2	3.15
<0.4	5.96	20.3	207	1,220	95.2	<0.02	0.41	<0.002	0.012	0.4	0.4	1.4	1.8	1.52
							10	1			10		50	30
50				500	250									

detection limit. Samples collected during this study having concentrations of 0.3 TU or less are interpreted to contain no modern water, whereas samples having more than 1 TU are interpreted to contain more than a small fraction of modern water. Apparent ^{3}H/^{3}He ages were computed for samples having concentrations of more than 0.3 TU.

In addition to ^{3}He derived from ^{3}H decay, groundwater also accumulates dissolved helium as it is produced from the radioactive decay of naturally occurring uranium- and thorium-series elements in aquifer solids ("crustal He") and from the upward advection/diffusion of primordial helium from the mantle ("mantle He"). Crustal and mantle helium are collectively referred to as "terrigenic He" (He$_{terr}$) (Solomon,

2000). Crustal- and mantle-sourced He are distinguishable by their relative abundance of ^{3}He and ^{4}He isotopes. These values are generally expressed as a ^{3}He/^{4}He (R) ratio relative to the atmospheric ^{3}He/^{4}He ratio (R$_a$). Because crustal He has an R/R$_a$ value of approximately 0.02 and mantle He has an R/R$_a$ value of approximately 10–30, the R/R$_a$ of a water sample provides information on the relative amount of crustal versus mantle sources of He$_{terr}$. Modern groundwater has an R/R$_a$ value approximately equal to 1, indicating that it contains atmospheric solubility concentrations of He. In most aquifers, crustal He makes up the majority of the He$_{terr}$. Where this is the case, the R/R$_a$ value of groundwater will fall below 1 as it acquires He$_{terr}$ from time spent in contact with the aquifer solids. Because He_{terr} concentrations generally increase with increasing residence time, dissolved $^4He_{terr}$ concentrations have been used as a semiquantitative tool for dating groundwater with ages from 10^3 to more than 10^6 years (Mazor and Bosch, 1992; Solomon, 2000). No attempts were made to accurately date groundwater in this study using $^4He_{terr}$, because crustal He_{terr} production rates are highly variable and substantial additional data would have been required to constrain these rates within the study area. However, Solomon (2000) reports average crustal ^4He production rates ranging from 0.28 to 2.4 μccSTP m^{-3}yr^{-1}. At these rates, groundwater should not acquire significant concentrations of $^4He_{terr}$ (more than about 2×10^{-8} ccSTP/g) until it has been in contact with the aquifer materials for more than about 1,000 years. Therefore, even without precise knowledge of local ^4He production rates, ^4He concentrations in excess of atmospheric solubility are useful as qualitative measures of groundwater age.

Carbon-14

Carbon-14 (^{14}C) is a naturally occurring radioactive isotope of carbon with a half-life of $5,730 \pm 40$ years that can be used to determine the apparent age of groundwater on time scales ranging from several hundred to more than 30,000 years. The method of radiocarbon dating is based on the radioactive decay of ^{14}C. In this study, the ^{14}C activity (its effective concentration) of dissolved inorganic carbon (DIC) was used to estimate the age of groundwater determined to be "premodern" by ^3H and ^3He. Kalin (2000) provides a comprehensive review of the radiocarbon groundwater dating method. Carbon-14 is produced in the upper atmosphere as cosmic rays react with atmospheric nitrogen-14 (^{14}N) to produce ^{14}C and hydrogen-1 (^1H). In the upper atmosphere, ^{14}C is rapidly oxidized to $^{14}CO_2$ which readily mixes into the lower atmosphere. Any material utilizing or reacting with atmospheric carbon dioxide (CO_2) (plants and water) has a ^{14}C activity equal to atmospheric ^{14}C while it is in equilibrium with the atmospheric carbon reservoir (Pearson and White, 1967). Carbon-14 activity is reported as percent modern carbon (pmc) and, by convention, the modern pre-1950 (prenuclear weapons testing) activity of atmospheric ^{14}C is 100 pmc. Carbon-14 generally enters the hydrologic cycle through any of four dominant pathways: (1) dissolution of atmospheric CO_2 into rain water and surface water, (2) plant respired CO_2 in the soil zone that dissolves into water, (3) CO_2 resulting from oxidation of organic material in the soil that dissolves into water, and (4) dissolution of mineral phases containing geologically young carbon.

The DIC in precipitation presumably has a ^{14}C activity in equilibrium with atmospheric CO_2. As precipitation infiltrates the subsurface, its ^{14}C activity is modified by carbon exchange with soil-zone CO_2 and minerals in the unsaturated zone until this infiltration enters the saturated zone. After entering the saturated zone, interaction with soil-zone carbon ceases and the ^{14}C in the DIC decays with time. The radiocarbon age of groundwater refers to the time that has elapsed since this water was isolated from carbon in the unsaturated zone.

In addition to radioactive decay, the ^{14}C activity of groundwater in the saturated zone can be affected by additions of and/or reactions with carbon-bearing minerals and organic phases. Four processes are of particular interest with respect to ^{14}C dating of groundwater: (1) dissolution of carbonate minerals such as limestone can increase the concentration of DIC having 0 pmc, thus decreasing the ^{14}C activity (Plummer and Sprinkle, 2001); (2) oxidation with older organic matter having 0 pmc can increase the concentration of DIC having low pmc, also decreasing the ^{14}C activity (Aravena and others, 1995); (3) sorption of calcium (Ca) and magnesium (Mg) ions to mineral surfaces may cause dissolution of carbonate minerals having 0 pmc, thus decreasing the ^{14}C activity (Plummer and others, 1990); and (4) carbonate mineral recrystallization (dissolution and subsequent precipitation of the same mass of carbonate mineral), which results in an isotope effect (Kendall and Caldwell, 1998) causing groundwater DIC to have a higher stable carbon isotope ratio ($\delta^{13}C$) and a lower ^{14}C activity (Parkhurst and Plummer, 1983).

These processes can greatly decrease the ^{14}C activity of groundwater. For example, in carbonate terrains such as the mountains surrounding Rush Valley, modern carbon in groundwater may be diluted with dissolved ^{14}C-*free* carbonate minerals to the extent that very young groundwater may have ^{14}C activities as low as 50 pmc (Clark and Fritz, 1997). Thus, adjustment is required to account for reaction effects on ^{14}C activity and obtain accurate radiocarbon ages. This is accomplished through a variety of models that attempt to quantify the processes described above to determine the ^{14}C activity of DIC derived from atmospheric CO_2 at the point of recharge—after the water passes through the unsaturated zone and prior to any reactions occurring within the aquifer. Several models exist to correct ^{14}C activity for the effects of the processes listed above. The most widely used formula-based models of this type are the Ingerson and Pearson (1964), Tamers (1975), and Fontes and Garnier (1979) models.

Ingerson and Pearson (1964) use a carbonate dissolution model to estimate initial ^{14}C activity (A_0) of groundwater DIC from $\delta^{13}C$ data for the inorganic carbon system, assuming that all DIC is derived from soil zone CO_2 and solid carbonates (Plummer and others, 1994). Disadvantages of the model are that it requires input that can be difficult to obtain and must often be assumed, such as the $\delta^{13}C$ of soil CO_2, and that it does not consider the effects of geochemical reactions other than mineral dissolution, particularly isotope exchange reactions. The Tamers (1975) model is a mass-balance model that considers only carbonate reaction with CO_2 gas and is based on chemical concentrations rather than $\delta^{13}C$ (Plummer and others, 1994). The dissolution of carbonate minerals dilutes ^{14}C activity by the reaction of dissolved CO_2 with solid carbonate to form bicarbonate (HCO_3^-). This model does not correct for the effects of isotope exchange. The Fontes and Garnier (1979) model is a hybrid of the Ingerson and Pearson (1964) model and the Tamers (1975) model, combining both chemical and isotopic data to correct for reaction effects on ^{14}C activity.

Oxygen-18 and Deuterium

The stable isotopes of water were used to better understand recharge sources to the groundwater basin. Most water molecules are made up of hydrogen (^{1}H) and oxygen-16 (^{16}O). However, some water molecules (less than 1 percent) contain the heavier isotopes of deuterium (^{2}H or D) and oxygen-18 (^{18}O). "Heavier" refers to the condition when there are additional neutrons in the nucleus of the hydrogen or oxygen atom, thereby increasing the mass or atomic weight of the water molecule.

Stable isotopes are analyzed by measuring the ratio of the heavier, less abundant isotope to the lighter, more abundant isotope and are reported as differences relative to a known standard. The isotope ratios are reported as delta (δ) values expressed as parts per thousand (permil). The δ value for an isotope ratio, R, is determined by:

$$\delta R = (R_{sample}/R_{standard} - 1) \times 1,000 \qquad (3)$$

where:

δR is the δ value for a specific isotope in the sample (^{2}H or ^{18}O),

R_{sample} is the ratio of the rare isotope to the common isotope for a specific element in the sample, and

$R_{standard}$ is the ratio of the rare isotope to the common isotope for the same element in the standard reference material. The reference standard used in this report is Vienna Standard Mean Ocean Water (VSMOW; Craig, 1961b; Coplen, 1994).

A positive δR value indicates that the sample is enriched in the heavier isotope with respect to the standard. A negative δR value indicates that the sample is depleted in the heavier isotope with respect to the standard. Heavier isotopes are more difficult to evaporate and easier to condense; for example, liquid contains more heavy isotopes than the vapor evaporated from the liquid. Because of this effect, water vapor in the atmosphere that condenses and falls out as precipitation will become progressively more depleted in the heavier isotopes at cooler temperatures and at higher altitudes. The proportional depletion of ^{2}H and ^{18}O results in isotopic compositions of precipitation (and groundwater sourced from precipitation) that plot along a trend referred to as a meteoric water line when the deuterium excess (δ^2H) is plotted versus the ^{18}O excess (δ^{18}O) (fig. 20).

Cooler (or high-altitude) precipitation values usually plot lower on the trend line and warmer (or low-altitude) precipitation values plot higher on the trend line. The trend line for worldwide precipitation defines the Global Meteoric Water Line (GMWL) and is described by the equation:

$$\delta^2 H = 8(\delta^{18}O) + d \qquad (4)$$

where:

d is defined as the ^{2}H excess (Dansgaard, 1964). The mean global value for d in freshwater is 10 (Craig, 1961a).

Depending on conditions and sources of precipitation, isotopic data from specific areas may plot along a trend line that is above or below the GMWL referred to as a local meteoric water line (LMWL). In addition to temperature, isotopic composition is also affected by evaporation, particularly during irrigation or from open-water bodies. Evaporation creates preferential enrichment in ^{18}O relative to ^{2}H, resulting in a shift from and a slope less than the LMWL or the GMWL. Groundwater with "evaporated" stable isotope compositions can often be identified as containing recharge from distinct sources such as lakes and irrigation canals.

Dissolved Gases

Dissolved-gas samples were collected and analyzed to evaluate groundwater recharge temperature (T_r, the temperature of recharging water as it crosses the water table) as an indicator of mountain versus valley recharge. A complete review of how dissolved noble gases are used as groundwater tracers is included in Stute and Schlosser (2000). Most noble gases that are dissolved in groundwater originate in the atmosphere. As recharging water enters the aquifer it becomes isolated from the atmosphere and the dissolved concentrations of most noble gases (Neon, Argon, Krypton, Xenon) become "fixed" according to their solubility relations to the temperature, pressure, and salinity conditions that existed at the water table at the time of recharge (Aeschbach-Hertig and others, 1999; Ballentine and Hall, 1999; Stute and Schlosser, 2000). These gases are generally nonreactive along flow paths in the subsurface, and their concentrations in groundwater, measured at points of discharge (wells and springs), provide a record of the physical conditions (temperature and pressure) that are related to the altitude of groundwater recharge. Although δ^{18}O and δ^2H are useful tracers for identifying groundwater that originated as mountain-derived precipitation, they cannot identify whether or not that precipitation recharge occurred "in-place" within the mountains or at valley altitudes, perhaps beneath losing streams entering the valley. In contrast, dissolved noble gas concentrations can be used to directly evaluate the relative contribution of mountain recharge to basin-fill aquifers (Manning and Solomon, 2003). For shallow water-table depths, recharge temperatures are generally within about 2°C of the mean annual air temperature at the location of recharge (Hill, 1990; Dominico and Schwartz, 1998). Mean annual air temperature decreases by about 10.5°C per mile of altitude gained in northern Utah (Hely and others, 1971); thus mountain recharge should be distinguishable from valley recharge by its colder T_r.

Measured concentrations of dissolved neon-20 (^{20}Ne), argon-40 (^{40}Ar), krypton-84 (^{84}Kr), and Xenon-129 (^{129}Xe) were used in a closed-system equilibration model (Aeschbach-Hertig and others, 2000; Kipfer and others, 2002) to calculate estimates of T_r, excess air (A_e), and a fractionation factor F, related to the partial dissolution of trapped air bubbles. The three unknown parameters (T_r, A_e, and F) were solved for by optimization of an overdetermined system of equations that relates them to the measured dissolved-gas concentrations in each sample.

To solve for the recharge parameters, the model requires that recharge altitude (H_r) be specified. Recharge altitude is a proxy for atmospheric pressure at the recharge location, which is required for the gas solubility calculations. Recharge altitude and T_r are highly correlated parameters, meaning that different combinations can produce nearly the same set of dissolved-gas concentrations. In areas of high topographic relief, it is generally not possible to know H_r ahead of time. Therefore, a range of T_r values was calculated using the largest possible range of H_r values. The minimum recharge altitude (H_{min}) for each sample was assumed to be the land-surface altitude at the sample site and the maximum recharge altitude (H_{max}) was assumed to be the highest altitude of the surface watershed in which the sample site is located. Maximum recharge altitude ranges from about 8,500 ft for wells in areas draining the Sheeprock Mountains to about 10,200 ft for wells in areas draining the Stansbury and Oquirrh Mountains. Using this method, the uncertainty in calculated values of T_r is approximately plus or minus 1°C (Manning and Solomon, 2003).

Results

Major Ions, Nutrients, and Selected Trace Metals

Analyses of major ions, nutrients, and selected trace metals were done on groundwater samples from 24 wells and one spring (table 8, fig. 15) to better define groundwater-source areas and flow paths and to describe general water-quality conditions in and immediately surrounding Rush Valley. The concentration of dissolved solids ranged from 181 to 1,590 mg/L (table 8, fig. 15). More than half of the sites sampled during this study have dissolved-solids concentrations that exceed the Environmental Protection Agency (EPA) secondary standard for drinking water of 500 mg/L (U.S. Environmental Protection Agency, 2009). Groundwater having the lowest dissolved-solids concentrations generally is found in the UCAU or the UBFAU, within or downgradient of the mountain recharge areas. The principal dissolved constituents in all samples with low dissolved-solids concentrations are Ca, Mg, and bicarbonate (HCO_3^-). Dissolved-solids concentration increases in the central part of the valley at the distal ends of the groundwater-flow paths. Water with high dissolved-solids concentrations is also found along the valley margins in areas that receive little recharge, where increased concentration is due mainly to greater amounts of sodium and chloride. This trend reflects the fact that groundwater has been in contact with aquifer materials for progressively longer periods as it moves from the mountains toward valley discharge areas or as it travels very slowly in areas receiving little recharge.

Geology also plays a role in the dissolved-solids concentration of groundwater. Several of the samples with the highest dissolved-solids concentrations are located in the southernmost part of southeastern Rush Valley, north of Vernon along Faust Creek, and along the central axis of the valley north of the groundwater divide (fig. 15). In each of these areas, groundwater likely moves through semiconsolidated rocks of the Tertiary Salt Lake Formation (LBFAU), indicating that it may be a strong source of dissolved solids. Previous studies have found similar trends. Hood and others (1969) reported groundwater having dissolved-solids concentrations of 767 to 2,180 mg/L north of Vernon along Faust Creek, and Enright (M. Enright, written commun., 1997) reported dissolved-solids concentrations in groundwater of greater than 1,000 mg/L in 19 wells (including 6 shallow wells with concentrations greater than 10,000 mg/L) along the central axis of the valley east of Clover.

Nutrient (nitrate plus nitrite) concentrations in all 25 samples were less than the EPA maximum contaminant level (MCL) of 10 mg/L. The four samples (sample sites 1, 11, 15, and 24; table 8) with the highest nutrient concentrations (≥ 3 mg/L) are from wells located in agricultural areas used for stock grazing or as cropland. Similar concentrations of nitrate in shallow wells near populated areas surrounding Clover and St. John have been reported (M. Enright, written commun., 1997, 1999).

Samples analyzed for arsenic (As) contained concentrations that exceed the EPA MCL of 10 µg/L at seven of 25 sample sites (table 8). Most waters with elevated As concentrations are from wells located south of the groundwater divide, north of Vernon and in southeastern Rush Valley (sample sites 5, 6, 8, 9, 11, 13; fig. 16). Elevated As concentrations were only found at one site in northern Rush Valley (sample site 23). Arsenic concentrations greater than the drinking-water threshold of 10 µg/L have been found in many aquifers worldwide (Levy and others, 1999; Karim, 2000; Planer-Friedrich and others, 2001; Ryu and others, 2002; Smedley and others 2002; Aiuppa and others, 2003). Volcanic rocks, such as those present in the subsurface and in outcrop (fig. 5) in the Oquirrh and Tintic Mountains and in the Vernon Hills, are commonly responsible for high concentrations of arsenic in groundwater (Smedley and Kinniburgh, 2002). Moreover, Casentini and others (2010) note that dissolved HCO_3^- enhances the leaching of As from volcanic rocks into groundwater. Waters with high As concentrations in Rush Valley were found in wells screened in the UBFAU or LBFAU (tables 8 and A1–2). Furthermore, As is generally associated with water that is thousands to tens of thousands of years old (groundwater age is discussed in subsequent sections) and appears to increase as a direct function of residence time. It is likely that the source of As in Rush Valley is from the mobilization of naturally occurring As (possibly enhanced by high concentrations of dissolved HCO_3^-) in alluvial sediments eroded from volcanic rock in the surrounding mountains.

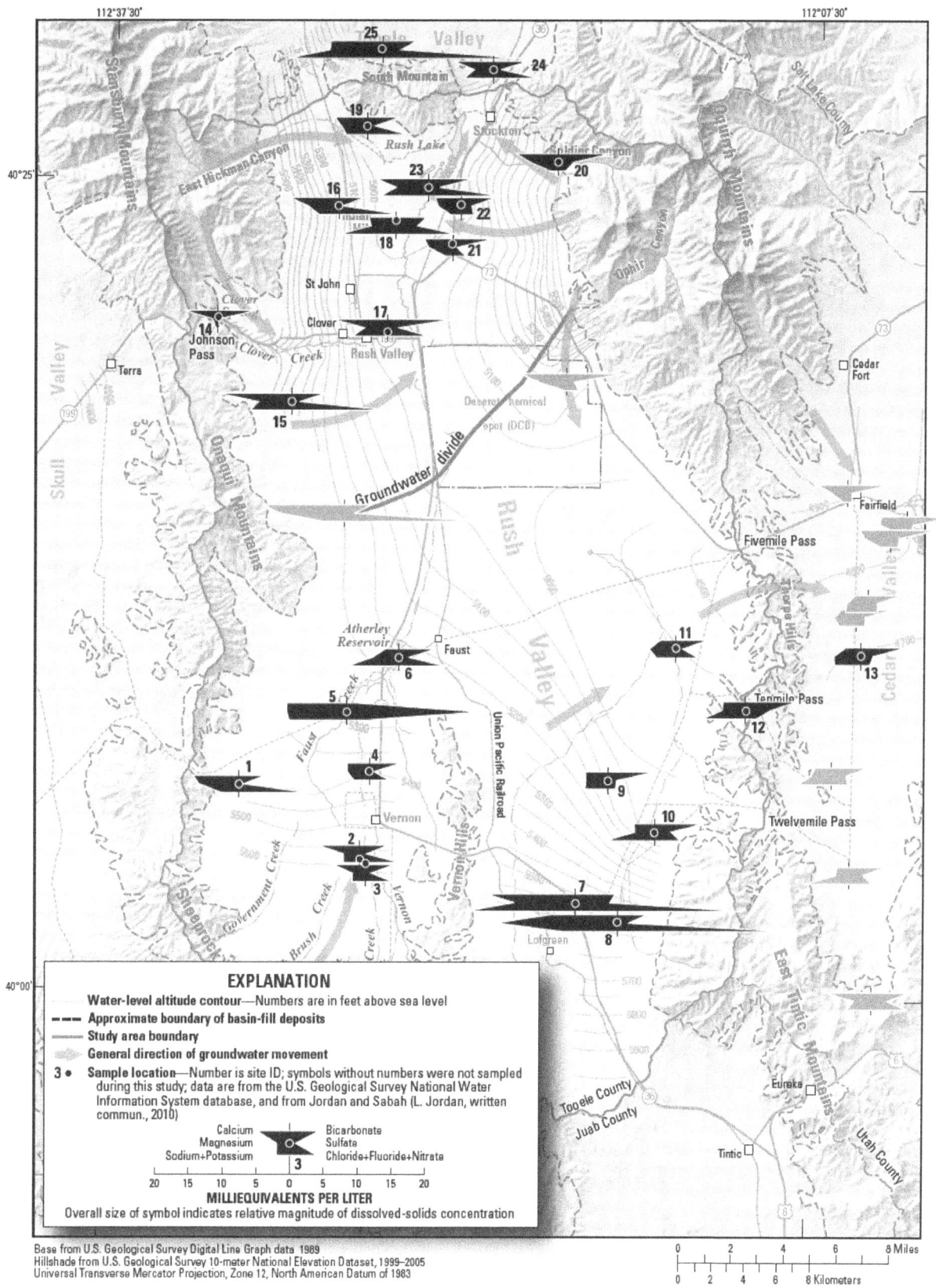

Figure 15. Location of sampled sites and Stiff diagrams showing major-ion composition of groundwater sampled during 2008–2010 in and around Rush Valley, Tooele County, Utah.

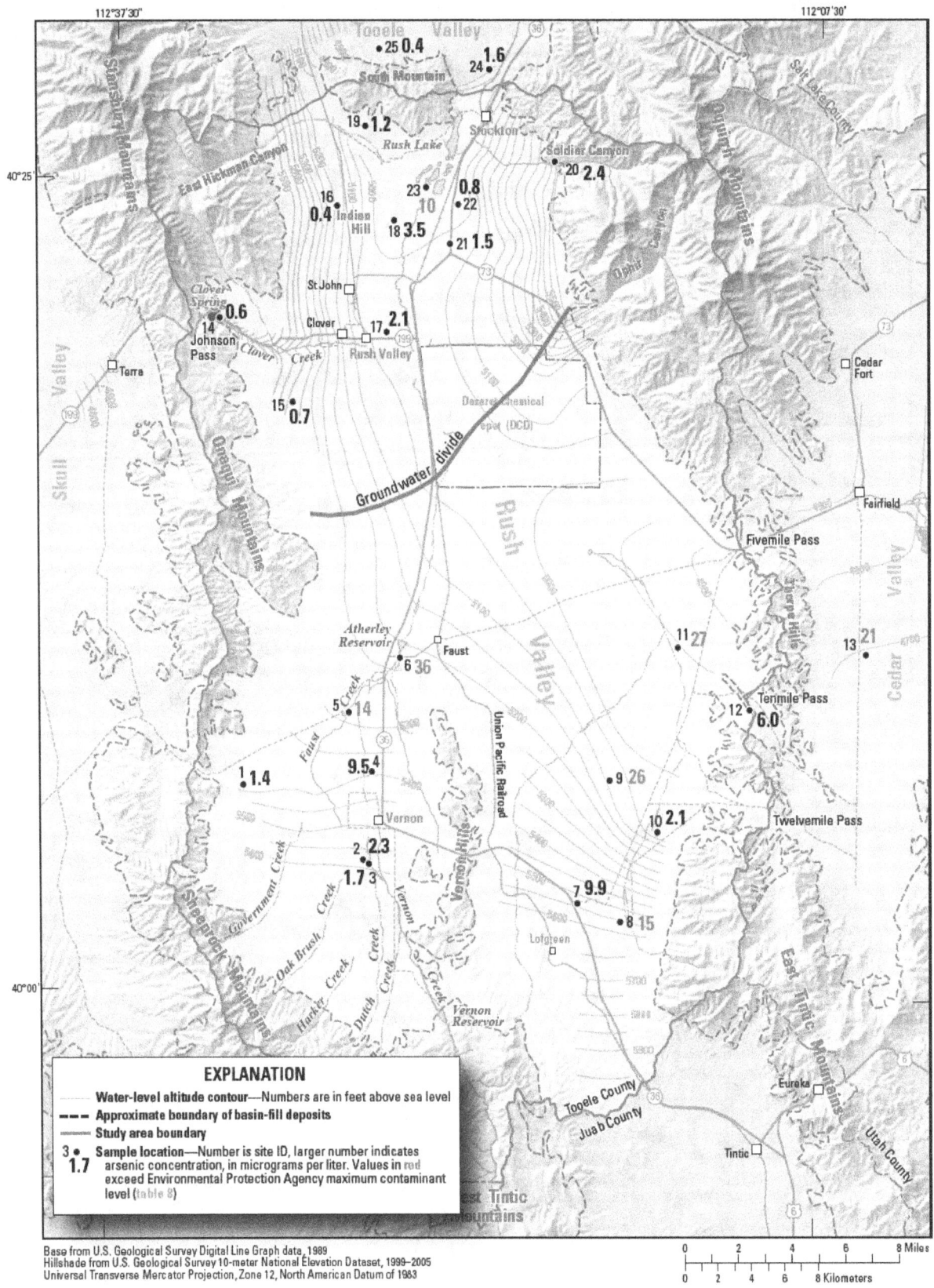

Figure 16.   Arsenic concentrations in groundwater sampled during 2008–09 in and around Rush Valley, Tooele County, Utah.

Tritium and Helium

Tritium concentrations greater than 0.3 TU were found in water from only 7 of 25 sample sites (fig. 17, table 9), indicating that most groundwater throughout Rush Valley was recharged prior to the mid-1950s. Concentrations ranged from less than detection (less than 0.1 TU) to 10 TU. With the exception of site 15 (0.4 TU), ^{3}H in all of the samples was clearly greater than or less than the modern water cutoff of 0.3 TU, allowing parts of the aquifer that have received recharge since about the mid-1950s to be identified. In northern Rush Valley, the presence of ^{3}H and ^{3}He$_{trit}$ indicate that modern water exists along the flow paths originating in the southern Stansbury Mountains along the Clover Creek drainage and in the Oquirrh Mountains in the vicinity of Soldier Canyon and north of Ophir Canyon. Similar ^{3}H concentrations were measured in water from Clover Springs (9.8 and 10 TU, site 14), discharging from the LBFAU in the southern Stansbury Mountains, and from a shallow UBFAU well near the town of Clover (9.2 TU, site 17). Tritium was less than detection (less than 0.1 TU) in shallow (less than 200 ft deep) UBFAU wells (sites 16 and 18) north of Clover and St. John in the western half of the valley (fig. 17, table 9). This pattern of modern groundwater focused south of the Stansbury Mountains and premodern water to the north supports the hypothesis that the geologic structure (steeply dipping bedding and a high-angle thrust fault) of the southern Stansbury Mountains (fig. 5) inhibits mountain recharge from flowing eastward, instead forcing it to flow south toward the Clover Creek drainage and possibly north along the axis of Rush Valley toward Tooele Valley. The ^{3}H concentration of 0.4 TU, detected in a shallow UBFAU well (site 15) south of Clover Creek, indicates that a minor fraction of modern recharge also may be moving through the aquifer south of Clover Creek, likely originating in the southern Stansbury or northern Onaqui Mountains.

Along the eastern half of northern Rush Valley, ^{3}H concentrations of 6.4 and 1.5 TU were detected in wells 20 and 21, respectively, indicating that at least a fraction of modern water moves through these areas. (fig. 17; table 9). Water from well 22 is premodern, with ^{3}H concentrations below detection. The direction of groundwater movement indicated by the water-level surface map (fig. 7) and the calcium-bicarbonate major-ion chemistry (fig. 15) indicate that all groundwater in this part of the valley originates as recharge through the UCAU in the Oquirrh Mountains to the east. The premodern water in well 22 could be the result of heterogeneous permeability in the basin fill, causing preferential flow to the north and south. However, the altitude of the well screens (land-surface altitude at the well minus the depth to the bottom of the open interval) for both of the wells with modern ^{3}H (sites 20 and 21) concentrations are above the altitude of Rush Lake (about 4,950 ft) while the well at site 22 is screened below this altitude. It is possible that as modern Oquirrh Mountain recharge moves westward into the valley, it is directed by low-permeability lakebed sediments to move primarily through the shallow UBFAU before discharging in springs along the eastern shore of Rush Lake (fig. 17).

Tritium concentrations greater than 0.3 TU were detected in water from only two wells in the Vernon area (sites 2 and 3). Both of these wells are complete in the UBFAU, upgradient (south) of Vernon (fig. 17). Wells to the west and north of this area (sites 1, 4, 5, and 6) have low ^{3}H values, indicating premodern water. The south-to-north direction of groundwater movement shown by the water-level surface map (fig. 17) and decreasing ^{3}H values along the same direction indicate that recharge in this area originates in the southeastern part of the Sheeprock Mountains. Furthermore, the low ^{3}H concentrations in groundwater downgradient of sites 2 and 3, screened in both fractured UCAU bedrock and the overlying UBFAU, indicate that much of the modern recharge moves through the shallow UBFAU and discharges in an area of seeps and springs west and north of Vernon (fig. 17). No modern groundwater was found in southeastern Rush Valley, indicating slow groundwater traveltimes in this area. This is not surprising in an area where the low-permeability LBFAU is ubiquitous (fig. 6B, section E–E') and recharge rates in the West Tintic Mountains and Vernon Hills are low (table 5).

Calculated apparent ^{3}H/^{3}He ages of modern waters range from about 3 to 35 years old (table 9). Apparent ^{3}H/^{3}He ages were not calculated for sites 9 or 17 because dissolved-gas samples used to determine ^{3}He$_{trit}$ were not collected due to gas bubbles observed in the well discharge. Tritium and dissolved-gas samples were collected at Clover Springs (site 14) during May (high flow) and September (base flow) of 2009; apparent ^{3}H/^{3}He ages were 3 and 3.5 years, respectively. This spring, located at an altitude of 5,900 ft in the southern Stansbury Mountains (fig. 15), had the youngest water of all sites sampled during this study; it discharges groundwater moving from the nearby recharge area through fractured carbonate rock of the LCAU. Apparent ages of the remaining modern groundwater samples (sites 2, 3, 15, 17, 20, and 21; fig. 17, table 9) were 16–35 years.

The relative mixtures of modern and premodern water of six samples from five sites having both measurable ^{3}H and ^{3}He$_{trit}$ (sites 3, 14, 15, 20, and 21) were examined by adding the nondecayed fraction of ^{3}H (measured ^{3}H) to the decay product (^{3}He$_{trit}$) to obtain a decay-corrected value referred to as "initial" ^{3}H. This value was compared to the historical record of ^{3}H concentration in precipitation at Salt Lake (fig. 18), located about 50 mi northeast of the study area. Mean annual ^{3}H concentrations in precipitation are derived from monthly data available for most of the period 1963–1984 for Salt Lake City (International Atomic Energy Agency, 2007). Mean annual ^{3}H concentrations prior to 1963 and after 1984 were estimated by correlation with ^{3}H concentrations in precipitation from Ottawa, Canada (International Atomic Energy Agency, 2007). Unmixed modern waters should plot close to the precipitation curve, having approximately the same "initial tritium" as precipitation at the time of recharge. Mixed waters should plot below the curve because the young fraction has been diluted with water that has very little ^{3}H or ^{3}He$_{trit}$; in other words mixed waters will have an apparent age of the young fraction because dilution with premodern water

Table 9.   Stable- and radio-isotope data used to estimate ages of groundwater sampled during 2008–09 in and around Rush Valley, Tooele County, Utah.

[Sample ID: see figure 15 for the location of sites sampled as part of this study and table A1–2 for physical information about the sampled well or spring. Hydrogeologic units and aquifer condition: UBFAU, upper basin-fill aquifer unit; LBFAU, lower basin-fill aquifer unit; UCAU, upper carbonate aquifer unit; LCAU, lower carbonate aquifer unit; C, confined; U, unconfined; ?, uncertain. Abbreviations: TU, tritium units; R, measured $^3He/^4He$ ratio; R_a, atmospheric $^3He/^4He$ ratio; ccSTP/g, cubic centimeters at standard temperature and pressure per gram of water; BP, before present; <, less than; —, not determined]

Sample ID	Hydrogeologic unit that well is completed in and aquifer condition	Sample date	$\delta^{18}O$, permil	δ^2H, permil	Tritium (^3H) and precision, TU[1]	R/R_a	Helium-4 (^4He), ccSTP/g	Terrigenic Helium-4 ($^4He_{terr}$), ccSTP/g[1]	Tritiogenic Helium-3 ($^3He_{trit}$), TU[1]	Carbon-14 (^{14}C) and precision, percent modern carbon	$\delta^{13}C$, permil	Apparent $^3H/He$ age, in years BP	Unadjusted ^{14}C age, in thousands of years BP	Adjusted ^{14}C age range, in thousands of years BP[2]
1	UBFAU, U	6/1/2009	-15.8	-120.3	<0.1+/-0.1	1.06	5.76E-08	5.70E-09	—	—	—	Premodern	—	—
2	UBFAU, U	6/3/2009	-15.5	-118.6	1.6+/-0.3	1.08	4.43E-08	8.32E-09	7.2	50.7+/-0.2	-10.1	30	Modern	Modern
3	UBFAU, U	6/3/2009	-15.5	-118.3	2.4+/-0.2	1.19	5.98E-08	3.83E-09	8.8	58.0+/-0.2	-11.0	28	Modern	Modern
4	UCAU, C	7/29/2008	-16.0	-121.1	<0.1+/-0.2	0.29	1.52E-07	1.07E-07	—	8.8+/-0.09	-7.6	Premodern	20	13–17
5	UBFAU and UCUA, C	6/5/2009	-16.9	-127.7	0.2+/-0.1	0.19	3.13E-07	2.62E-07	—	1.9+/-0.04	-8.8	Premodern	33	30–33
6	UBFAU, C	5/29/2009	-17.0	-128.6	<0.1+/-0.1	0.35	1.29E-07	5.57E-08	—	—	—	Premodern	—	—
7	LBFAU, ?	4/28/2009	-15.4	-119.2	0.2+/-0.1	0.72	8.13E-08	2.80E-08	—	—	—	Premodern	—	—
8	LBFAU, ?	5/6/2009	-16.3	-125.1	<0.1+/-0.1	0.71	7.03E-08	2.13E-08	—	8.0+/-0.07	-9.3	Premodern	21	15–19
8[3]	LBFAU, ?	5/6/2009	-16.3	-125.0	<0.1+/-0.1	0.70	7.00E-08	2.08E-08	—	8.0+/-0.09	-9.3	Premodern	21	15–19
9[4]	UBFAU, U	5/15/2009	-17.1	-130.9	0.1+/-0.1	—	—	—	—	7.1+/-0.08	-8.2	Premodern	22	15–20
10	LBFAU, U	4/28/2009	-18.0	-138.6	<0.1+/-0.1	0.11	5.56E-07	5.06E-07	—	0.4+/-0.02	-7.4	Premodern	46	42–45
11	UBFAU, U	4/22/2009	-16.3	-125.9	<0.1+/-0.1	0.75	6.34E-08	1.48E-08	—	8.7+/-0.09	-6.9	Premodern	20	11–18
12	LCAU, ?	4/22/2009	-17.4	-133.2	<0.1+/-0.1	0.05	1.24E-06	1.19E-06	—	2.2+/-0.05	-5.1	Premodern	32	22–32
13	?, U	5/15/2009	-16.8	-126.9	<0.1+/-0.1	0.62	7.81E-08	3.32E-08	—	13.1+/-0.1	-6.9	Premodern	17	7.5–13
14	LCAU, C	5/12/2009	-16.3	-121.4	10+/-1.7	1.02	4.00E-08	3.10E-09	2.2	—	—	3.5	—	—
14	LCAU, C	9/25/2009	-16.5	-121.2	9.8+/-0.5	1.02	3.90E-08	2.36E-09	1.8	69.9+/-0.3	-9.2	3	—	—
15	UBFAU, U	6/2/2009	-15.9	-123.3	0.4+/-0.1	1.01	5.18E-08	2.91E-09	2.5	—	—	35	Modern	Modern
16	UBFAU, U	6/2/2009	-16.8	-127.5	<0.1+/-0.1	0.74	6.48E-08	1.96E-08	—	—	—	Premodern	—	—
17[4]	UBFAU, C	5/26/2009	-16.0	-119.5	9.2+/-0.4	—	—	—	—	—	—	Modern[5]	—	—
18	UBFAU, C	5/6/2009	-17.1	-129.5	<0.1+/-0.1	0.38	2.19E-07	1.80E-07	—	4.8+/-0.06	-4.6	Premodern	25	13–24
19	UCAU, ?	7/29/2008	-16.8	-128.3	0.1+/-0.2	0.70	9.44E-08	3.06E-08	—	4.4+/-0.07	-7.5	Premodern	26	19–25
20	UCAU, ?	6/5/2009	-16.6	-123.1	6.4+/-0.5	1.08	8.46E-08	1.65E-08	11.3	54.6+/-0.2	-8.9	18	Modern	Modern
21	UBFAU, C	7/29/2008	-16.8	-126.0	1.5+/-0.1	1.00	4.62E-08	3.57E-09	2.3	—	—	16	—	—
22	UBFAU, U	5/29/2009	-17.2	-129.4	<0.1+/-0.1	0.57	1.21E-07	6.23E-08	—	—	—	Premodern	—	—
23	UBFAU, C	5/12/2009	-16.8	-127.4	0.2+/-0.1	0.31	1.08E-06	1.03E-06	—	29.4+/-0.2	-8.5	Premodern	10	1.6–6.2[6]
24	UBFAU, U	5/26/2009	-15.1	-118.2	<0.1+/-0.1	0.46	1.14E-07	7.01E-08	—	13.5+/-0.3	-10.0	Premodern	17	11–13
25	UCAU, ?	6/2/2009	-16.1	-124.8	0.2+/-0.1	0.18	7.90E-07	7.46E-07	—	9.2+/-0.08	-7.1	Premodern	20	11–18

[1]Interpreted value derived using the Closed-Equilibrium dissolved gas model (Aeschbach-Hertig and others, 2000; Kipfer and others, 2002)

[2]Range represents minimum to maximum radiocarbon age using the formula-based correction models of Ingerson and Pearson (1964), Tamers (1975), and Fontes and Garnier (1979) converted to calendar years before present using the Fairbanks 0107 calibration curve (Fairbanks and others, 2005)

[3]Replicate sample included for quality control

[4]Dissolved-gas sample not collected due to presence of gas bubbles in well discharge

[5]Apparent $^3H/He$ age was not calculated because dissolved-gas sample was not collected. Water determined to be modern (< 60 years old) based on presence of ^3H only

[6]Measured ^{14}C activity is suspect due to potential contamination with modern carbon and because elevated dissolved ^4He concentration indicates that this water may be much older

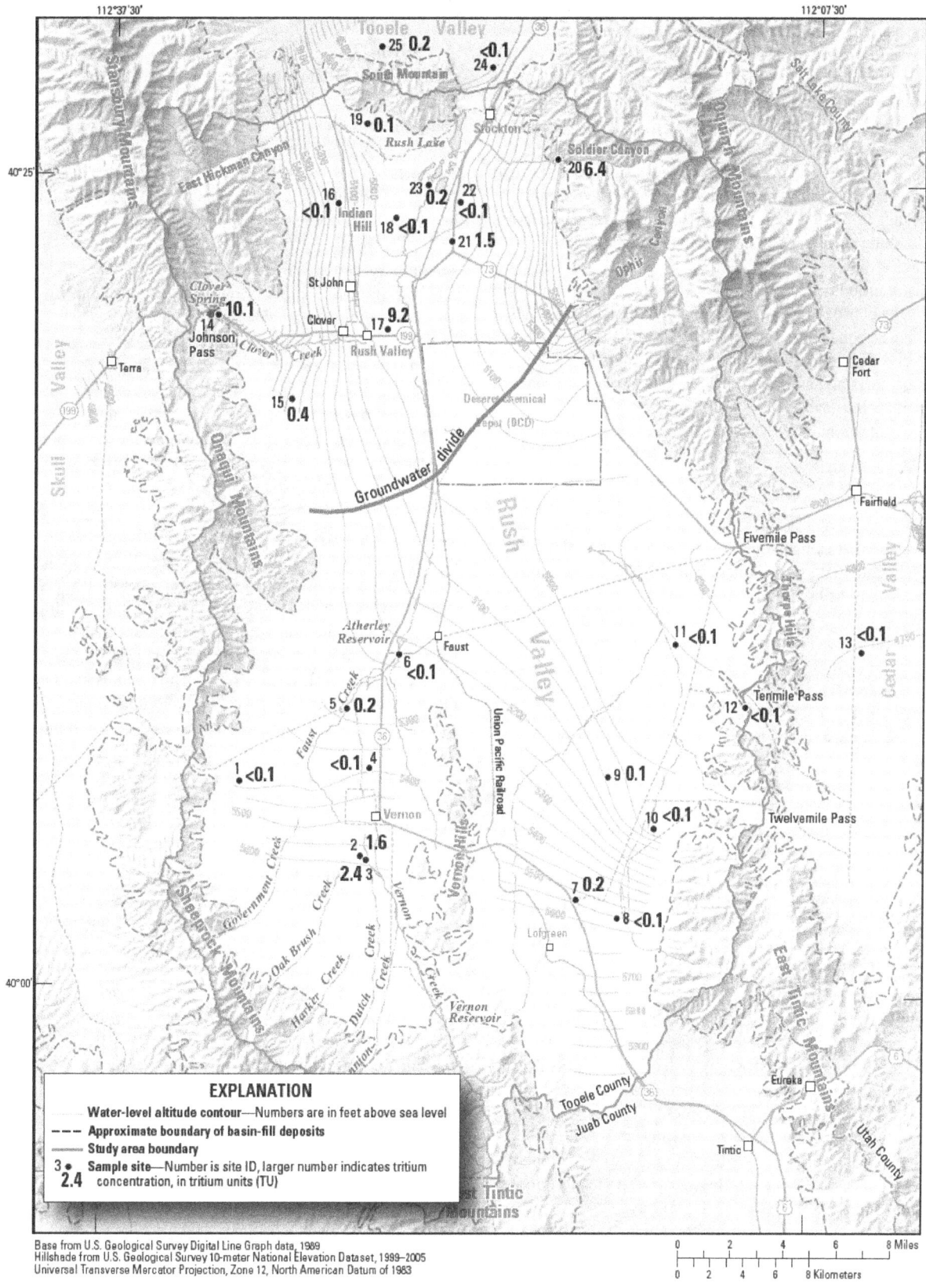

Figure 17. Tritium concentrations in groundwater sampled during 2008–09 in and around Rush Valley, Tooele County, Utah.

will not significantly alter the ^{3}H/^{3}He$_{trit}$ ratio. Samples from two carbonate aquifer sites, a well (site 20) completed in the UCAU at the mouth of Soldier Canyon and Clover Spring (site 14; collected during both peak flow and baseflow), plot very near the precipitation curve and appear to be composed of nearly all modern water. Samples from three wells completed in the UBFAU on the western and eastern sides of northern Rush Valley (sites 15 and 21, respectively) and in the southern part of the Vernon area (site 3) plot below the precipitation curve, indicating that these samples are mixtures of modern and premodern water (fig.18). The modern/premodern mixtures at the three basin-fill sites indicate that the wells are capturing groundwater with a wide range of ages as coalescing flow paths approach localized discharge areas. The samples from the carbonate aquifer sites may also be a mixture of water of different ages; however, these sites are located near recharge areas and any mixture is likely composed of modern water of various ages.

Concentrations of ^{4}H$_{eterr}$ were measured in groundwater from 23 of 25 sample locations and ranged from 2.36x10^{-9} to 1.19x10^{-6} ccSTP/g (table 9). As discussed in the methods section of this report, ^{4}He$_{terr}$ was not used to assign specific ages to groundwater in this study, but rather as an indicator of relative age that can be compared to ^{3}H/^{3}He and radiocarbon ages. The rate that groundwater acquires ^{4}He$_{terr}$ from aquifer materials may be highly variable from one location to another across the study area. Even so, a sample containing more than about 2.0x10^{-8} ccSTP/g should have a minimum mean age of more than 1,000 years. In general, samples that contain modern water (^{3}H > 0.3 TU) have low ^{4}He$_{terr}$ concentrations, between about 2.4x10^{-9} and 8.3x10^{-9} ccSTP/g, as would be expected. One exception to this pattern is sample site 20, where water with 6.4 TU contains about 1.7x10^{-8} ccSTP/g of ^{4}He, which is high enough to suggest the presence of much older water. Site 20 is a 620-ft-deep municipal well with a 200-ft screened interval in fractured carbonate rock. Although the previously discussed comparison of "initial ^{3}H" to the historical record of ^{3}H in precipitation indicates that this sample contains mostly modern water, it is likely that this deep bedrock well captures a small fraction of older water causing the sample to have elevated ^{4}He$_{terr}$. Most premodern samples (^{3}H < 0.3 TU) have more than 2.0x10^{-8} ccSTP/g of ^{4}He, indicating not only that they are older than about 60 years (the limit of the ^{3}H/^{3}He dating method) but that they likely contain large fractions of water that are thousands or more years old. One notable exception is at sample site 1, where ^{3}H is below detection but ^{4}He$_{terr}$ is only 5.7x10^{-9} ccSTP/g, which is within the range of the other modern samples. Although water from site 1 is too old to be dated using ^{3}H (pre- mid-1950s), the low ^{4}He$_{terr}$ implies that it is closer in age to the group of modern water samples than to most of the group of premodern water samples. Samples from sites 12 and 23 have the highest ^{4}He$_{terr}$ concentrations of all waters sampled during this study (1.19x10^{-6} and 1.03x10^{-6} ccSTP/g, respectively). These concentrations are nearly two orders of magnitude above concentrations of ^{4}He found in modern waters, indicating that these sites may contain some of the oldest waters in the study area.

Carbon-14

Carbon-14 activity measured from DIC in groundwater samples from 17 sites within and immediately surrounding Rush Valley ranged from 0.4 to 69.9 pmc (fig. 19; table 9). Samples from four of the sites (2, 3, 14, and 20) containing water determined to be modern using ^{3}H/^{3}He dating had ^{14}C activities of 50.7, 58.0, 69.9, and 54.6 pmc, respectively. Sites 14 and 20, located near the Stansbury and Oquirrh Mountain recharge areas, discharge groundwater from the LCAU and UCAU, respectively. Groundwater at sites 2 and 3 in the southern Vernon area discharge from the UBFAU, down gradient of the Sheeprock Mountain recharge area.

Of the remaining 13 sites, one of the samples (site 23) had an intermediate ^{14}C activity of 29.4 pmc, and the remaining 12 samples had very low ^{14}C activities, ranging from 0.4 to 13.5 pmc, indicating that groundwater throughout much of the study area may be thousands to tens of thousands of years old.

Radiocarbon ages were calculated for the 13 samples with ^{14}C data that were determined to be premodern based on ^{3}H concentrations of less than 0.3 TU (sites 4, 5, 8, 9, 10, 11, 12, 13, 18, 19, 23, 24, and 25). Table 9 lists unadjusted and adjusted radiocarbon ages for these samples. Radiocarbon ages are not shown for sites 2, 3, 14, and 20 because tritium concentrations indicate they contain significant fractions of water that are less than 60 years old. The range of adjusted ages represents the minimum and maximum ages determined using the formula-based adjustment models of Ingerson and Pearson (1964), Tamers (1975), and Fontes and Garnier (1979). The adjusted ages were converted to calendar years before present (BP) using the Fairbanks 0107 calibration curve (Fairbanks and others, 2005). Adjusted radiocarbon ages for each of these models are compared to unadjusted ages in table A1–3 in Appendix 1. These radiocarbon-age adjustment models require ^{14}C and δ^{13}C values of soil zone CO_2 and carbonate minerals that are available to react with or add to the DIC of the groundwater. In all cases, ^{14}C activity was assumed to be 100 pmc and soil zone CO_2 was assumed to have a δ^{13}C value of -23.3 permil, the average isotopic value reported for soil zone CO_2 in the nearby Wasatch Mountains (Cerling and others, 1991). Carbonate minerals were assumed to have 0 pmc ^{14}C because of their age (middle Cambrian to Permian) and a δ^{13}C value of 0, approximately the worldwide average for marine limestones (Keith and Weber, 1964).

Unadjusted radiocarbon ages range from 10,000 to 46,000 years (table 9). The largest age adjustments (difference between unadjusted and adjusted "minimum" radiocarbon ages) range from 3,000 to 12,000 years, giving adjusted "minimum" ages of 1,600 to 42,000 years. Eleven of the 13 sample sites have minimum adjusted radiocarbon ages of 11,000 years or more, indicating that groundwater moves very slowly through large parts of Rush Valley and that these areas may have received more recharge prior to the Holocene, when the climate was cooler and wetter. The remaining two sites (13 and 23) are located along the distal ends of their respective flow paths. As such, they would be expected to have older

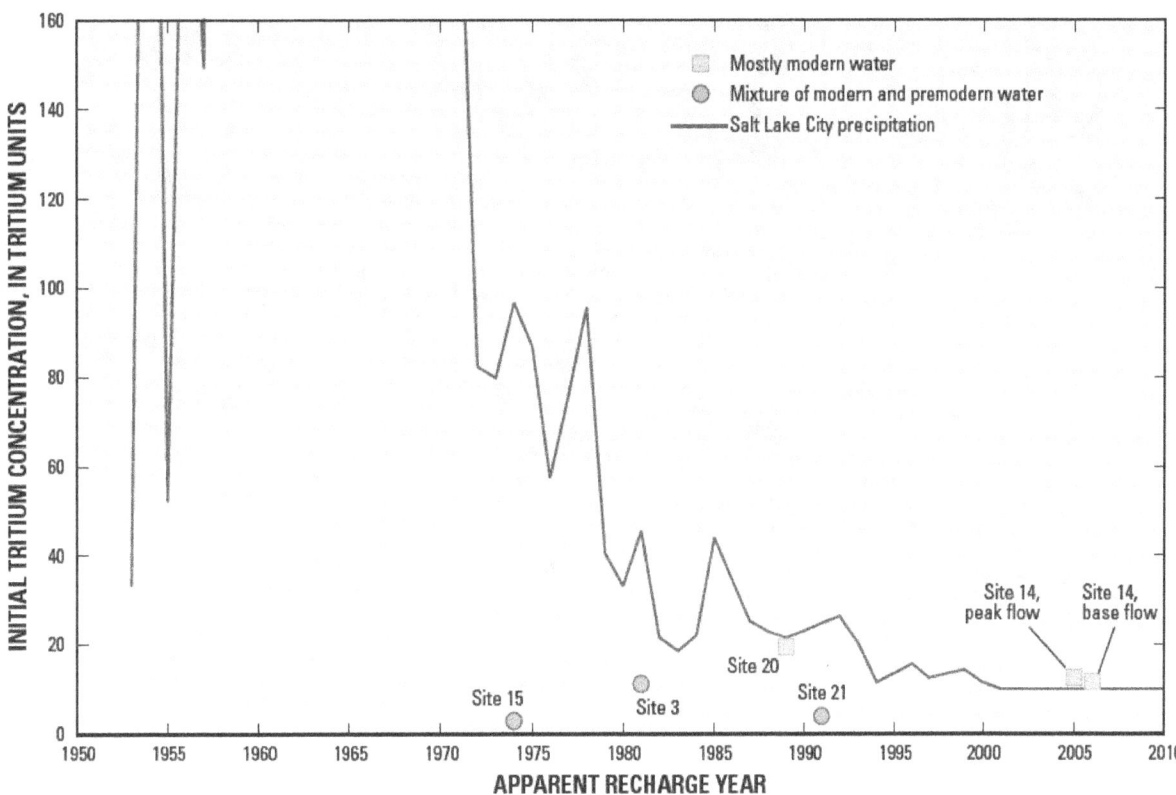

Figure 18. Tritium concentrations in precipitation and the relation between initial tritium concentration and apparent recharge year for groundwater sampled during 2009 in and around Rush Valley, Tooele County, Utah.

rather than younger radiocarbon ages compared to the sites discussed above. Site 13 is a 198-ft deep well in western Cedar Valley with an adjusted minimum radiocarbon age of 7,500 years. Major-ion chemistry indicates that this may be a mixture of groundwater moving through southeastern Rush Valley (for example sites 9 and 11 on fig. 15) with water recharged in the southern Oquirrh Mountains that moves southeastward into Cedar Valley. A mixture of old water from southeastern Rush Valley with modern recharge from the southern Oquirrh Mountains would explain the apparent younger age of this water relative to many of the valley sites sampled. Site 23, a flowing well located in the discharge area of northern Rush Valley near the center of Rush Lake, has an adjusted minimum radiocarbon age of only 1,600 years. Water from this well also contained a small amount of ^{3}H (0.2 TU) and a very high concentration of ^{4}He$_{terr}$ (1.03x10^{-6} ccSTP/g), indicating that it is probably a mixture of water of very different ages. However, because site 23 is a large-diameter flowing well with an open casing filled with dense algal growth, the DIC content of this water may be contaminated with dissolved modern CO_2 from algal respiration and decomposition, artificially resulting in an erroneously young radiocarbon age. Given this scenario, and the very high ^{4}He$_{terr}$ found in this sample, a large fraction of the groundwater from this site is likely to be much older than indicated by its radiocarbon age.

Oxygen-18 and Deuterium

Stable-isotope ratios of all groundwater samples collected during this study plot along (or slightly below) the Utah meteoric water line (UMWL) derived by Kendall and Coplen (2001), indicating that they are waters of meteoric origin (fig. 20, table 9). The sample waters are classified as modern and premodern on figure 20. The general age classifications are included for the purpose of identifying climatic effects on the stable-isotope compositions. Samples classified as "modern" are those that contain detectable ^{3}H and are therefore less than about 60 years old; "premodern" samples contain no detectable ^{3}H and/or have adjusted radiocarbon ages ranging from about 1,600 to 40,000 years old (table 9). Premodern samples with minimum radiocarbon ages of more than 11,000 years are separately identified as Pleistocene on figure 20.

The Rush Valley groundwater samples are plotted along with stable-isotope compositions of water samples collected from across north-central Utah that include mountain and mountain front streams, valley streams, and water from thermal springs (fig. 20). Stable-isotope compositions for 29 north-central Utah valley streams range from -79 to -125 permil and from -8.6 to -16.8 permil for δ^2H and δ^{18}O, respectively. This wide range of values occurs when stream water sourced from precipitation falling at a range of altitudes, that is then held in lakes and reservoirs, undergoes varying

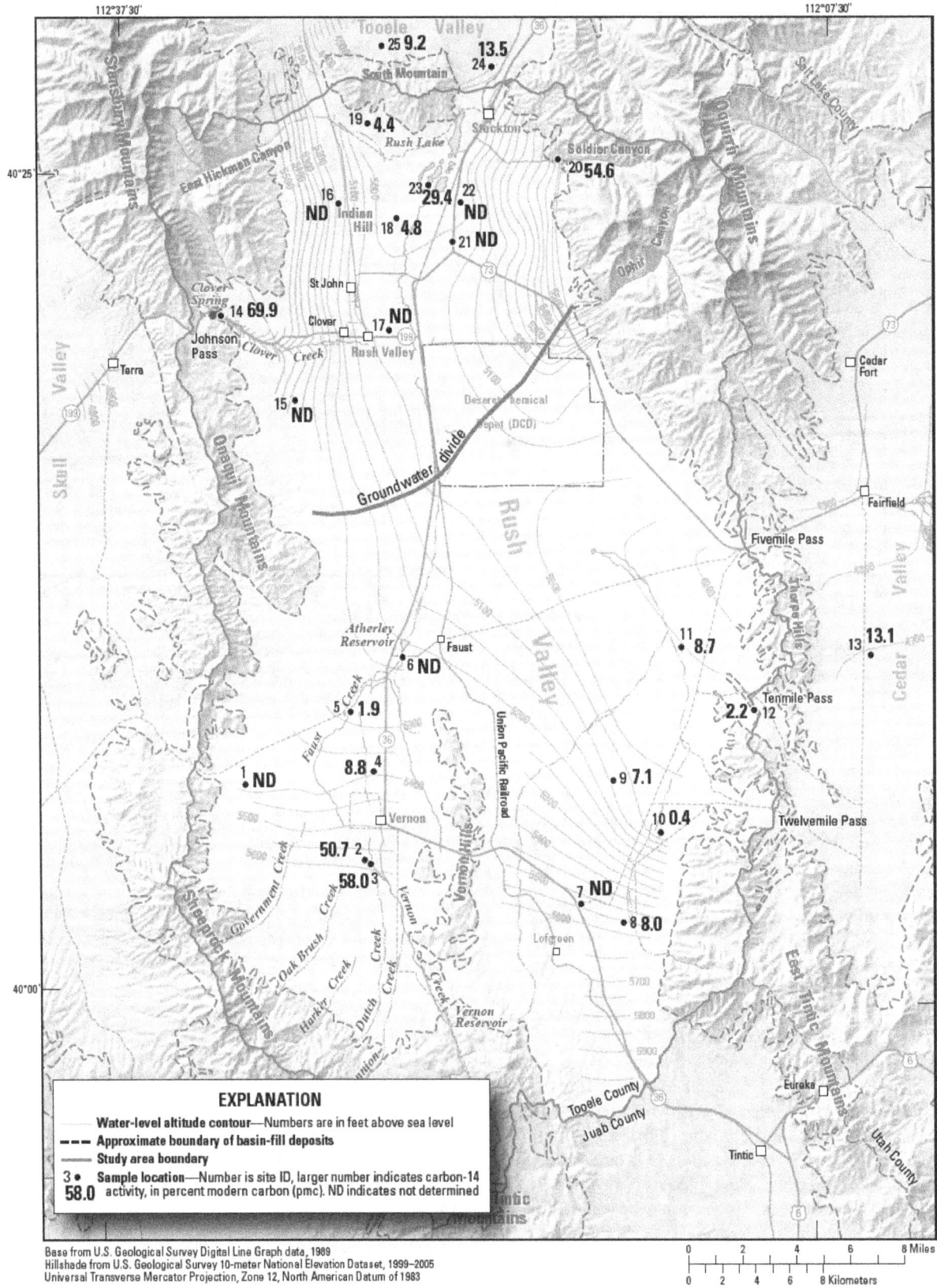

Figure 19. Carbon-14 concentrations in groundwater sampled during 2008–09 in and around Rush Valley, Tooele County, Utah.

degrees of evaporative enrichment and mixing. Stable-isotope compositions for 35 north-central Utah mountain and mountain-front streams ranged from -118 to -128 permil and from -16.0 to -17.2 permil for δ^2H and $\delta^{18}O$, respectively. These samples represent integrated (well-mixed) modern mountain precipitation that has not been evaporated to any significant degree. Modern groundwater in Rush Valley that is sourced from mountain precipitation should have similar stable-isotope compositions regardless of whether it originated as in-place mountain recharge or as stream-loss recharge near the mountain fronts. Samples sourced from precipitation falling at lower altitudes should be isotopically heavier (have less negative values) and plot higher and farther to the right along the UMWL.

Only about 6 of 25 Rush Valley samples are significantly heavier ($\delta^{18}O$ values greater than -16.0 permil) than modern high-altitude precipitation (fig. 20, table 9). The majority of $\delta^{18}O$ values are less than or equal to -16.0 permil, indicating that very little recharge originates from low-altitude precipitation, which agrees with BCM estimates of little to no recharge occuring in valley areas where precipitation is generally less than 14 in./yr. Five of the six samples (sites 1, 2, 3, 7, and 15, fig. 15; table 9) that are more enriched than -16.0 permil are from UBAFU wells where recharge likely originates in the Onaquai, Sheeprock, or West Tintic Mountains. All three of these areas have lower average altitudes than many north-central Utah mountain ranges. For this reason, recharge is expected to be more isotopically enriched than most mountain precipitation for the region represented by the "mountain and mountain-front stream" waters in figure 20. The remaining sample that is more enriched than -16.0 permil is from site 24, located north of the Stockton Bar, where small amounts of groundwater are thought to move through the subsurface from Rush Valley to Tooele Valley. Hood and others (1969) suggested that water from Rush Lake seeps into the ground and moves slowly northward toward Tooele Valley. If this occurs, the isotopic composition of water from site 24 can be explained as a mixture of the isotopically enriched seepage from Rush Lake with deeper groundwater in the surrounding UBAFU.

Two characteristics of the isotopic composition of groundwater samples from Rush Valley are notable when grouped as modern and premodern waters (fig. 20): (1) several of the premodern waters are more depleted (plot farther down and to the left on the UMWL) than any of the mountain and mountain-front stream samples representing modern mountain precipitation and, (2) although most modern Rush Valley waters plot very close to the UMWL, premodern samples all plot below the line. Both of these patterns indicate that premodern groundwater recharged under different climatic conditions.

Smith and others (2002) used stable-isotope compositions of Great Basin waters to identify Pleistocene recharge (more than about 10,000 years old). They noted that Pleistocene recharge could contain waters both isotopically lighter and heavier than modern precipitation, depending on the source of recharge. Lighter waters will have been recharged from glacial meltwater or directly from precipitation. Heavier waters will

have been recharged from lakes that formed during the pluvial (wetter) period in ice-free areas (for example Lake Bonneville) and may have somewhat evaporated isotopic compositions, resulting in a shift below and to the right of the UMWL. Samples of the heavier waters recharged during the Pleistocene will not be identifiable using stable isotopes because they may appear to be evaporatively enriched modern precipitation. However, Great Basin groundwater with isotopic compositions lighter than any high-altitude present-day precipitation can be classified as Pleistocene recharge. On the basis of this earlier work, the two most isotopically depleted groundwater samples from Rush Valley (sample sites 10 and 12, fig. 15; fig. 20) can be classified as having been recharged during the Pleistocene, a time when conditions were cooler and wetter than at the present time. These samples are from southeastern Rush Valley, where recharge rates and aquifer permeabilities are low. Although no other premodern samples can be classified as Pleistocene based solely on being more depleted than modern mountain precipitation, the other premodern groundwater samples from Rush Valley all plot below the UMWL, which is consistent with somewhat evaporated lake-water recharge that could have occurred during a pluvial period of the Pleistocene.

Dissolved Gases

Dissolved-gas concentrations are presented for 23 of the 25 sample sites and dissolved-gas recharge temperatures (T_r) are presented for 21 of the 25 sample sites in table 10. Samples were not collected at two of the sites (9 and 17) where water from the well had visible air bubbles, because the results would not have been reliable. Two samples were collected at sites 8, a well (one a replicate for quality assurance) and 14, a spring (one during high flow and another during baseflow). Measured gas concentrations were unreliable in samples from sites 2 and 11 due to either an ice blockage during extraction or an air bubble in the sample tube, and recharge temperatures were not derived for these samples.

The range of possible T_r values calculated for each of the 23 sites is shown in figure 21, in which the left and right points for each sample represent T_{rmin} and T_{rmax}, respectively. Because T_r represents the temperature of the water table at the location of recharge, T_r values are compared to valley water-table temperatures to identify areas where groundwater is composed of mountain rather than valley recharge.

Modern (or Holocene) mountain recharge should have T_r values cooler than the temperature of the water table in the valley. Measured groundwater temperatures from 14 valley wells in Rush Valley, assumed to represent the water-table temperature, ranged from 10.0 to 14.6°C with an average value of 12.6°C (fig. 21). At each of these wells, the water temperature was measured under static conditions where the water table was less than 165 ft or, if under pumping conditions, at wells screened shallower than 165 ft.

Minimum dissolved-gas recharge temperatures (T_{rmin}) range from 0.8 to 16.2°C and maximum dissolved-gas recharge temperatures (T_{rmax}) range from 3.2 to 19.4°C (fig. 21, table 10). Nearly all T_r values are outside of the range

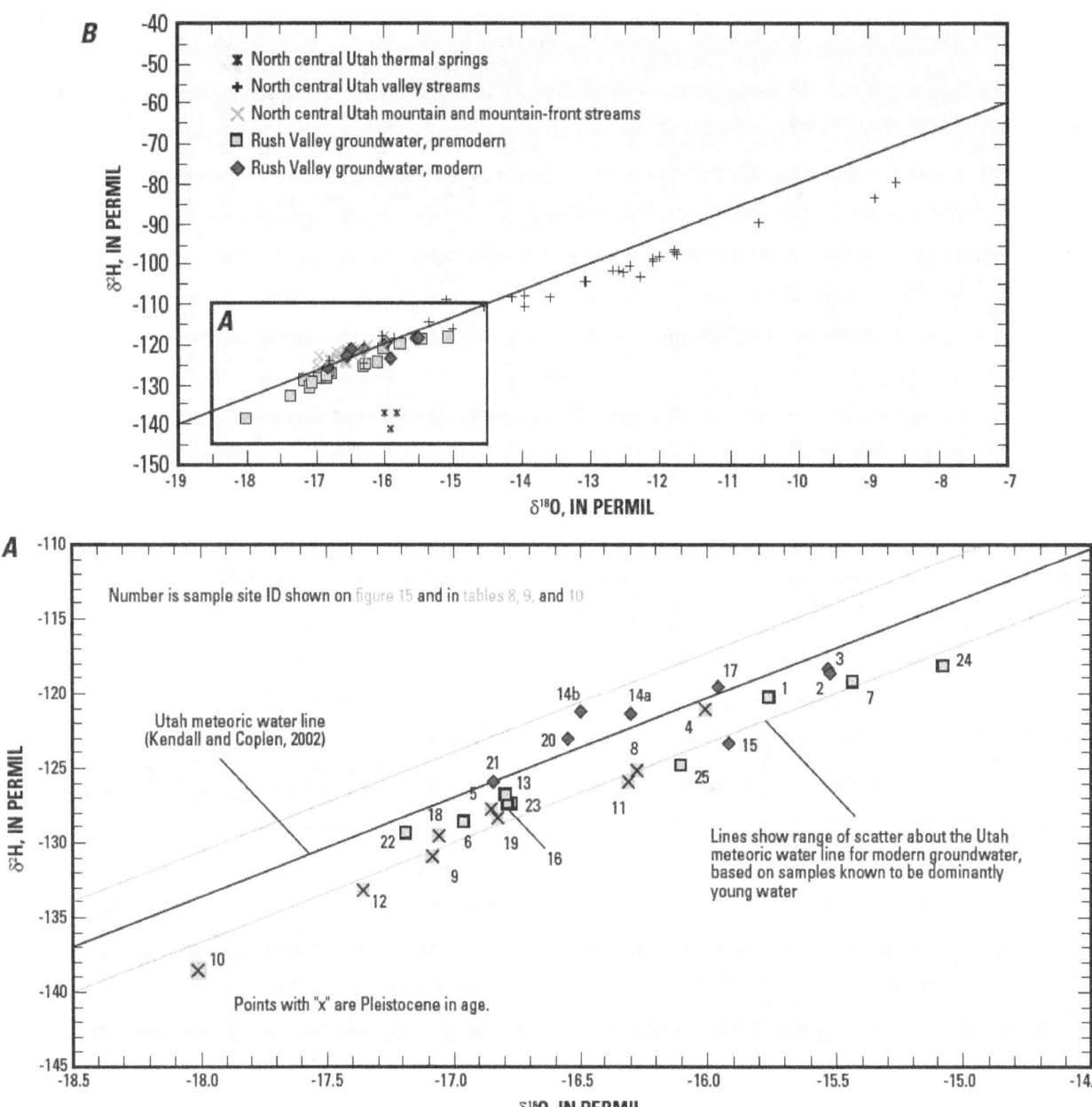

Figure 20. Stable-isotope values of groundwater samples collected *A*, during this study (2009) in and around Rush Valley compared to *B*, ground- and surface-water samples collected throughout north-central Utah (1981–2004).

of measured valley water-table temperatures. Dissolved-gas recharge temperatures are generally cooler than valley recharge for 19 of the samples and warmer for two of the samples (sites 12 and 19). The average recharge temperature (T_{ravg}, the mid-point for each sample displayed on fig. 21) is calculated assuming that the recharge altitude (H_r) is closer to the median altitude within a watershed. All values of T_{ravg} from this "cool" subset of 19 sample sites are less than the coolest measured valley water-table temperature (10.0°C) and most (14 of 19) are less than 7.6°C, the measured discharge temperature at Clover Spring (site 14), which likely contains

100 percent mountain recharge. Some of the samples likely represent integrated mixtures of groundwater-flow paths with a range of recharge temperatures. However, two-end-member mixing between mountain recharge (assumed $T_r = 3°C$ because much of it is assumed to be infiltrating snowmelt at high altitudes) and valley recharge (assumed $T_r = 12.6°C$, the average valley water-table temperature) requires 52-percent mountain recharge to obtain a sample with an apparent $T_r = 7.6$, indicating that even mixed samples contain mostly mountain recharge. Samples with minimum radiocarbon ages greater than 11,000 years old (Pleistocene in age) are an

Table 10. Dissolved-gas and recharge temperature data for groundwater sampled during 2008–09 in and around Rush Valley, Tooele County, Utah.

[Sample ID: see figure 15 for the location of sites sampled as part of this study and table A1–2 for physical information about the sampled well or spring Hydrogeologic unit and aquifer condition: UBFAU, upper basin-fill aquifer unit; LBFAU, lower basin-fill aquifer unit; UCAU, upper carbonate aquifer unit; LCAU, lower carbonate aquifer unit; C, confined; U, unconfined, ?, uncertain Dissolved-gas sample collection method: CT, copper tube; DS, diffusion sampler; NC, not collected mmHg, millimeters of mercury; ccSTP/g, cubic centimeters at standard temperature and pressure per gram of water; °C, degrees Celsius; ?, unknown; —, not determined]

Sample ID	Hydrogeologic unit that well is completed in and aquifer condition	Sample date	Dissolved-gas sample collection method	Barometric pressure, mm Hg	Dissolved-gas pressure, mm Hg	Neon-20 (^{20}Ne), ccSTP/g	Argon-40 (^{40}Ar), ccSTP/g	Krypton-84 (^{84}Kr), ccSTP/g	Xenon-129 (^{129}Xe), ccSTP/g	Recharge temperature range, °C[1]
1	UBFAU, U	6/1/2009	CT	623	768	1.96E-07	3.48E-04	4.67E-08	2.94E-09	6.5–9.1
2[2]	UBFAU, U	6/3/2009	CT	626	668	1.12E-07	1 28E-04	1.15E-08	5.86E-10	—
3	UBFAU, U	6/3/2009	CT	628	743	2.08E-07	3 57E-04	4.47E-08	3.06E-09	6.1–8.9
4	UCAU, C	7/29/2008	CT	—	642	1.80E-07	3 51E-04	4.57E-08	3.18E-09	4.6–7.5
5	UBFAU and UCUA, C	6/5/2009	DS	622	710	2.05E-07	4 10E-04	5.62E-08	3.73E-09	2.1–3.2
6	UBFAU, C	5/29/2009	CT	636	637	1.61E-07	4.03E-04	4.81E-08	3.64E-09	2.0–4.1
7	LBFAU, ?	4/28/2009	CT	620	826	2.02E-07	3 53E-04	4.08E-08	2.97E-09	8.7–11.0
8	LBFAU, ?	5/6/2009	CT	622	671	1.88E-07	3.63E-04	4.97E-08	3.28E-09	3.0–5.8
8[3]	LBFAU, ?	5/6/2009	CT	—	—	1.91E-07	3.70E-04	4.83E-08	3.29E-09	3.9–6.4
9[4]	UBFAU, U	5/15/2009	NC	633	681	—	—	—	—	—
10	LBFAU, U	4/28/2009	CT	625	768	1.94E-07	3 91E-04	4.92E-08	3.31E-09	4.6–8.0
11[2]	UBFAU, U	4/22/2009	CT	633	677	1.72E-07	3 21E-04	3.54E-08	2.20E-09	—
12	LCAU, ?	4/22/2009	CT	—	707	1.83E-07	3 28E-04	4.46E-08	2.41E-09	16.2–18.3
13	?, U	5/15/2009	CT	643	646	1.78E-07	3 54E-04	4.55E-08	3.14E-09	3.9–8.5
14	LCAU, C	5/12/2009	DS	612	629	1.50E-07	3.43E-04	4.74E-08	3.35E-09	1.2–5.8
14	LCAU, C	9/25/2009	DS	620	617	1.49E-07	3 34E-04	4.85E-08	3.33E-09	0.8–6.1
15	UBFAU, U	6/2/2009	CT	627	673	1.87E-07	3 51E-04	4.48E-08	3.07E-09	4.4–8.7
16	UBFAU, U	6/2/2009	DS	631	634	1.78E-07	3 59E-04	4.77E-08	3.10E-09	4.3–8.7
17[4]	UBFAU, C	5/26/2009	NC	637	—	—	—	—	—	—
18	UBFAU, C	5/6/2009	DS	638	613	1.49E-07	2 96E-04	4.14E-08	2.63E-09	6.6–12.4
19	UCAU, ?	7/29/2008	CT	—	663	2.27E-07	3 15E-04	3.79E-08	2.18E-09	14.6–19.4
20	UCAU, ?	6/5/2009	CT	612	782	2.55E-07	4.08E-04	5.10E-08	3.35E-09	2.8–6.7
21	UBFAU, C	7/29/2008	CT	—	—	1.65E-07	3 25E-04	4.33E-08	2.84E-09	6.1–10.6
22	UBFAU, U	5/29/2009	CT	639	713	2.16E-07	3.60E-04	4.76E-08	3.09E-09	3.8–8.8
23	UBFAU, C	5/12/2009	DS	635	699	1.92E-07	3.70E-04	4.94E-08	3.33E-09	2.0–6.5
24	UBFAU, U	5/26/2009	DS	641	671	1.70E-07	3 34E-04	4.37E-08	3.04E-09	4.3–9.4
25	UCAU, ?	6/2/2009	CT	635	709	1.75E-07	3 20E-04	3.92E-08	3.22E-09	5.1–10.7

[1]Interpreted value derived using the Closed-Equilibrium dissolved-gas model (Aeschbach-Hertig and others, 2000; Kipfer and others, 2002)

[2]Measured dissolved-gas concentrations were unreliable due to either incomplete laboratory extraction of gases or an air bubble in the sample tube causing gas stripping Reliable recharge temperature could not be calculated

[3]Replicate sample for quality control

[4]Dissolved-gas sample not collected due to presence of gas bubbles in well discharge Presence of ^{3}H indicates water is modern

exception because they may contain water that recharged at valley altitudes under cooler climatic conditions.

The two samples (sites 12 and 19) with T_r warmer than measured valley water-table temperatures are from deep wells in fractured bedrock of the LCAU and UCAU. Both of these wells are located on the periphery of Rush Valley (fig. 15) in areas where the depth to water is greater than 300 ft and temperatures near the water-table are warmer than about 22°C, considerably warmer than the mean valley water-table temperature of 12.6°C. Deep water tables are significantly

influenced by the geothermal gradient (the rate at which the Earth's temperature increases with depth), resulting in water-table temperatures above the mean annual air temperature. These premodern waters likely recharged where the depth to water is great enough that the ground temperature at the water table (thus, T_r) is controlled by the local geothermal gradient rather than the average annual air temperature.

Dissolved-gas recharge temperature data provide another line of evidence to support a conceptual model where most groundwater within Rush Valley is composed of either modern

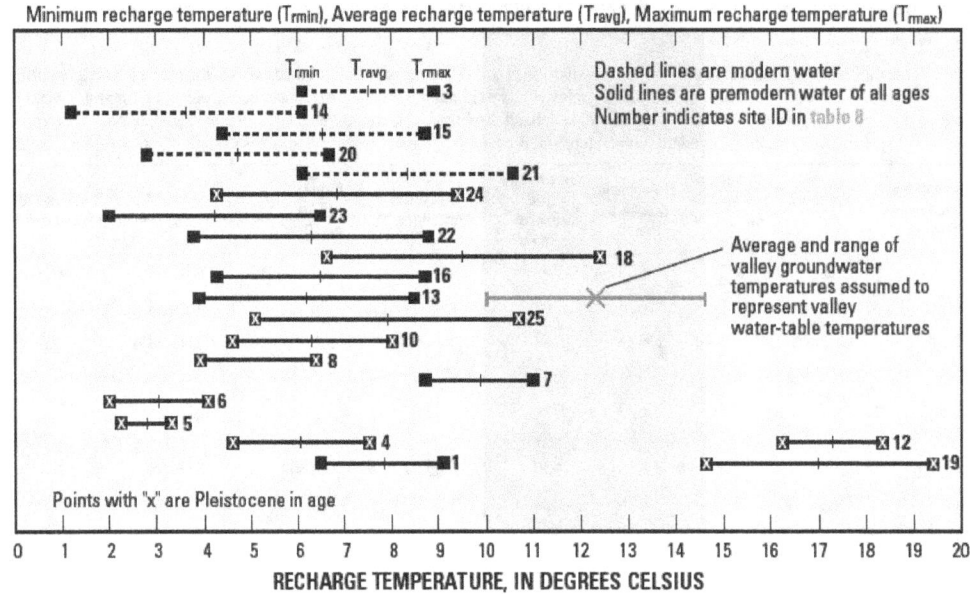

Figure 21.   Dissolved-gas recharge temperatures of groundwater samples collected during 2008–09 in and around Rush Valley, Tooele County, Utah.

(or Holocene) recharge that originated as infiltration of precipitation within the mountains or of Pleistocene recharge that may have occurred at valley altitudes during a cooler climate. These data also agree well with previous estimates (Hood and others, 1969) showing recharge in the mountains as the largest source of water to the basin-fill aquifers.

Discussion

In many ways Rush Valley is similar to other Basin and Range groundwater systems. Most recharge originates as infiltration of precipitation that falls on the high-altitude mountains surrounding the valley. Where the mountains are composed of permeable bedrock, recharge enters the aquifer system as direct infiltration of precipitation in the mountains and moves toward the valley and into the basin fill. This groundwater then moves toward discharge areas at lower altitudes within the valley. However, geochemical and environmental tracer data collected during this study provide additional information about the patterns of recharge and discharge, flow pathways, and groundwater residence times in different parts of the study area.

Nearly all groundwater sampled for ^{3}H and ^{14}C during this study falls into two distinct groups: (1) samples containing modern water with apparent ^{3}H/^{3}He ages of only decades and (2) samples composed of much older water, generally with apparent ages of more than 11,000 years. All waters in the first group have ^{3}H concentrations greater than 0.3 TU and ^{14}C activities of more than 50 pmc. Waters in the second group have ^{3}H concentrations less than or equal to 0.3 TU and, with the exception of sample site 23, ^{14}C activities of less than 14 pmc (fig. 22). As previously mentioned, the intermediate ^{14}C activity (29.4 pmc) and minimum radiocarbon age (1,600

years) of water from site 23 may be misleading due to the possibility that the DIC content of this water sample was contaminated with dissolved modern CO_2. Not all sites were sampled for ^{14}C; it is possible, therefore, that premodern waters of intermediate age (less than 11,000 years old) exist. However, the apparent bimodal distribution of available age data indicates that such intermediate ages are scarce. These data also indicate that parts of the groundwater system are more active, receiving modern recharge and circulating groundwater on timescales of decades, while other areas are less active, receiving little recharge and having substantially slower groundwater traveltimes.

When making groundwater-age-based interpretations, the applicable age range for the tracers being used must be considered as well as the possibility that samples likely contain mixtures of water of different ages. Waters older than about 60 and younger than about 300–500 years are too old to be dated using ^{3}H, given its short half-life (12.32 years) and too young to be reliably dated using radiocarbon because of its long half-life (5,730 years). There currently are no methods to evaluate groundwater ages between about 60 and 500 years old. Nonetheless, water that is less than a few hundred years old would still be characterized as part of the "young" group in the bimodal age distribution of Rush Valley groundwater.

Most groundwater samples are composed of mixtures of water of different ages. This is because springs generally occur in discharge areas where groundwater-flow paths converge and wells, screened over intervals of tens to hundreds of feet, capture groundwater from multiple flow paths. In a groundwater system that contained a continuum of water from Pleistocene to modern in age, where the oldest water occurred at the end of long or slow flow paths, one would expect to obtain intermediate mixed samples with apparent ages

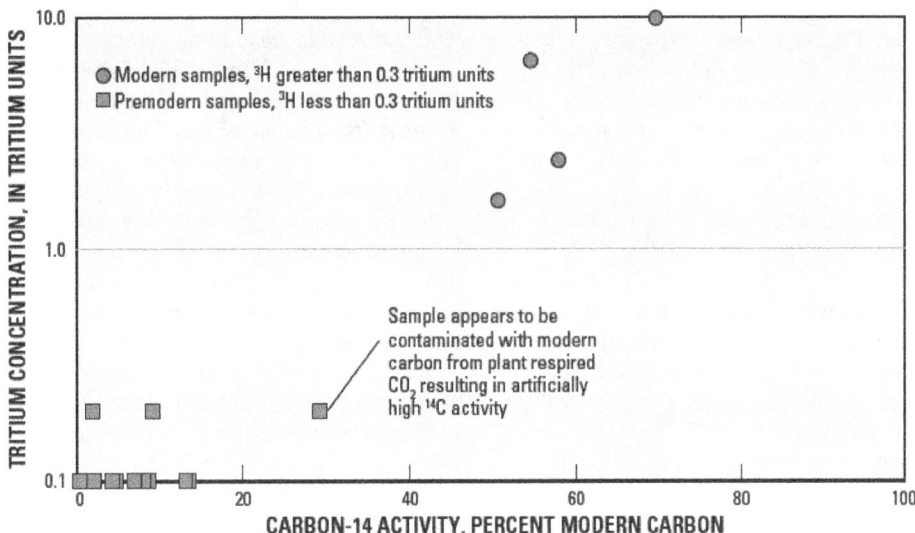

Figure 22. Tritium concentrations versus carbon-14 activity measured in groundwater sampled during 2009 in and around Rush Valley, Tooele County, Utah.

spanning the entire range from modern to more than 10,000 years old. However, this is not the case for much of Rush Valley, where most groundwater appears to be either relatively modern or much older.

Figure 23 displays the interpreted ages of groundwater samples. Apparent ^{3}H/^{3}He ages are shown for samples containing modern water and adjusted minimum radiocarbon ages are shown for premodern samples at locations where ages could be calculated. Available data were insufficient for calculating apparent groundwater ages at 7 of 25 sample sites (sites 1, 6, 7, 16, 17, 22, and 23). These waters are classified only as "modern" or "premodern" based on their ^{3}H concentrations. For example, no apparent age is given for sample site 17 (fig. 23) because a noble-gas sample was not obtained; thus, a ^{3}H/^{3}He age could not be calculated. Instead, water from this site is classified simply as modern based on the sample having a ^{3}H concentration of 9.2 TU. Interpretations about the relative ages of the remaining waters in this group (sites 1, 6, 7, 16, 22, and 23) are described in the following paragraphs.

Sample site 23, a flowing well located in the central discharge area of northern Rush Valley, is classified as premodern due to its low ^{3}H concentration (0.2 TU). The ^{14}C activity of this sample (29.4 pmc) is suspect and it has a very high ^{4}He$_{\text{terr}}$ concentration (1.03x10^{-6} ccSTP/g). Although the minimum adjusted radiocarbon age calculated for water from site 23 is only 1,600 years, the ^{4}He$_{\text{terr}}$ concentration is among the highest of any measured during this study, indicating that this sample contains some of the oldest water in the study area. End-member mixing calculations were made using ^{14}C, ^{4}He$_{\text{terr}}$, and ^{3}H concentrations from representative young and old waters to examine whether or not mixing could be responsible for this combination of measured values. Site 21

was chosen to represent the young component (^{14}C, assumed 55 pmc, ^{4}He$_{\text{terr}}$ = 3.57x10^{-9}ccSTP/g, and ^{3}H =1.5 TU) because it contains modern water from a nearby shallow well, and site 12 was chosen to represent the old component (^{14}C = 2.2 pmc, ^{4}He$_{\text{terr}}$ = 1.19x10^{-6}ccSTP/g, and ^{3}H = <0.1 TU) because it is the only other sample site where water had similarly high ^{4}He$_{\text{terr}}$. A mixture of 10-percent young and 90-percent old water results in ^{4}He$_{\text{terr}}$ and ^{3}H values of 1.08x10^{-6}ccSTP/g and 0.15 TU, respectively, which are very similar to those observed at site 23 listed above. However, this mixture yields a ^{14}C activity of only 9 pmc, which is much lower than the measured value of 29.4 pmc. Even if the old water component of this mixture had a ^{14}C activity of 14 pmc, similar to the highest value measured in the remaining premodern waters, the resulting water would still only have a ^{14}C activity of 19 pmc. On the basis of this reasoning, it seems likely that water from site 23 is a mixture of mostly very old water with a small fraction of young water and that the DIC content is contaminated with carbon from dissolved modern CO$_2$.

Sites 1, 6, 7, 16, and 22 were not sampled for ^{14}C, but all have ^{3}H concentrations less than 0.3 TU and are classified as premodern (table 9, fig. 23). Site 1 has no detectable ^{3}H but is likely not much older than other waters classified as modern, because it has very little ^{4}He$_{\text{terr}}$ (5.70x10-9 ccSTP/g). Sample site 7 has low ^{3}H (0.2 TU) and elevated ^{4}He$_{\text{terr}}$ (2.80x10^{-8} ccSTP/g), indicating that this sample may contain a mixture of water that is slightly older than modern and water that is probably thousands of years old. Sample sites 6, 16, and 22 contain elevated ^{4}He$_{\text{terr}}$ concentrations (5.57x10^{-8}, 1.96x10^{-8}, and 6.23x10^{-8} ccSTP/g, respectively), indicating that these waters are at least thousands of years old and possibly much older. Based largely on these groundwater age interpretations, the principal basin-fill aquifer system in Rush Valley is conceptualized as having (1) areas of more active groundwater movement (indicated as the "approximate extent of modern or nearly-modern groundwater" on figure 23) that exist along relatively focused flow paths between mountain recharge and valley discharge locations and (2) areas of much less active groundwater movement that surround and underlie the more active areas.

The first of three more active areas of groundwater movement is on the western side of northern Rush Valley where modern groundwater (sites 14, 15, and 17) is found in the UBFAU along flow paths that generally parallel the Clover Creek drainage. This area is fed by recharge occurring in the southern Stansbury Mountains that appears to be directed southeastward along the trend of the Big Hollow Thrust Fault and a steeply dipping section of the USCU that parallels this fault to enter the UBFAU in the vicinity of Clover Creek (fig. 5, section C–C'). Groundwater moves eastward through this area and then northward toward an area of discharge south of Rush Lake. The second area of more active flow is on the eastern side of northern Rush Valley, north of Ophir Canyon, where modern groundwater (sites 20 and 21) moves from the mountains through the UBFAU to an area of discharge along the eastern side of Rush Lake (fig. 23). Premodern water was found at site 22, a 236-ft deep well, that is at least several thousand and possibly more than 10,000 years old based on its elevated $^4He_{terr}$ concentration (6.23×10^{-8} ccSTP/g). Site 22 is located about 1.5 mi north of site 21 and is screened approximately 150 ft deeper in the UBFAU, indicating that the majority of groundwater in this area moves around site 22 or that movement is restricted to only the shallow UBFAU. Although no wells were available to sample south of Ophir Canyon and north of Fivemile Pass, the gentle southeast hydraulic gradient that parallels the mountain front indicates that recharge to this area is probably minimal and that groundwater movement is slow. As previously discussed, the lack of recharge to this area is attributed to the geologic structure and orientation of bedrock confining units (USCU) in the southern Oquirrh Mountains that direct recharge away from Rush Valley and toward Cedar Valley to the east. The third area of more active groundwater movement is in the southern part of the Vernon area where modern (sites 2 and 3) or nearly modern (site 1) groundwater moves northward from the Sheeprock Mountains through the UBFAU to an area of discharge located north of Vernon and along Faust Creek. Water from wells screened at greater depths (sites 4 and 5) and located downgradient (site 6) of this area appears to be much older, indicating that, similar to northern Rush Valley, most modern recharge circulates through the shallow UBFAU.

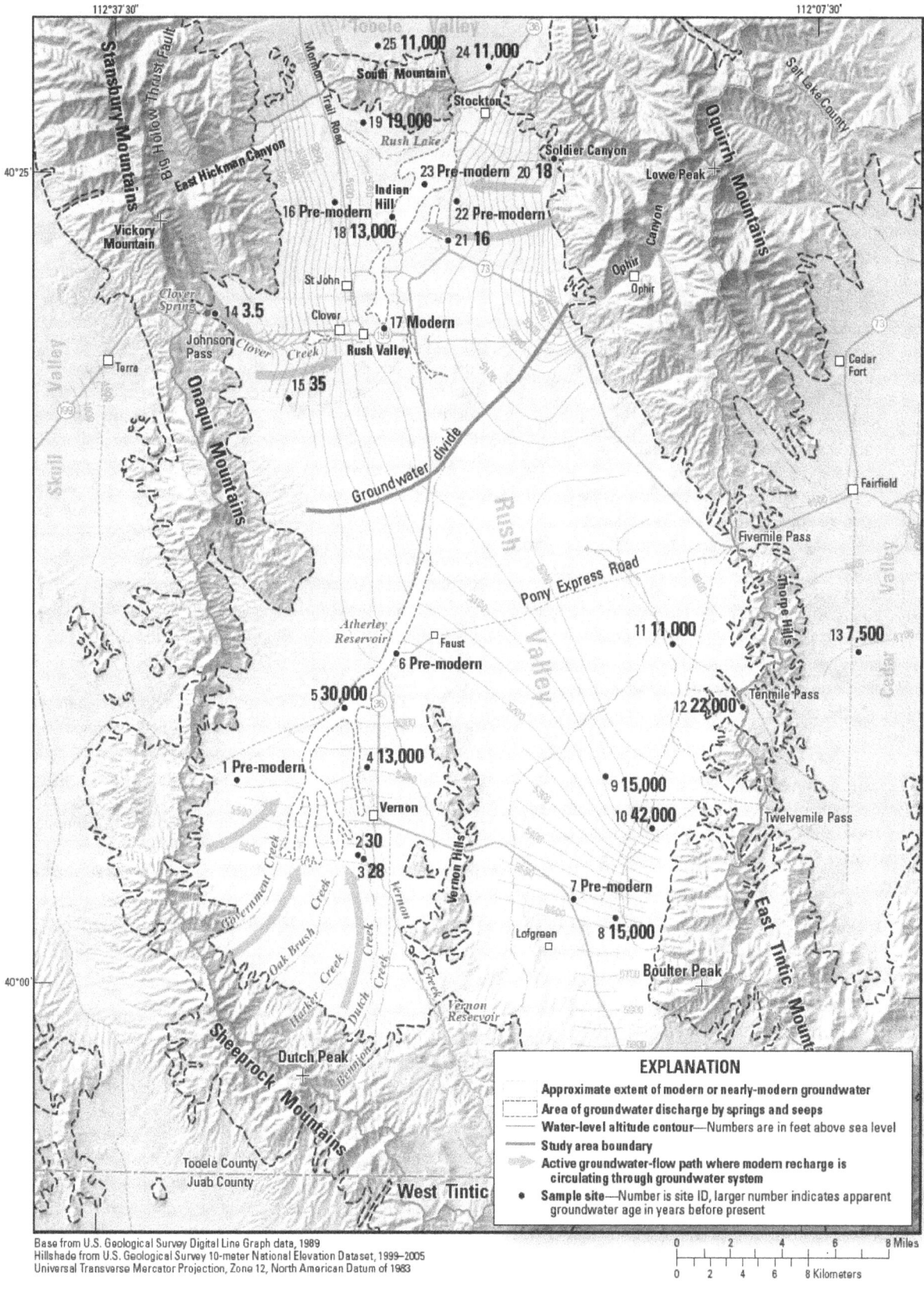

Figure 23. Distribution of apparent groundwater age, interpreted age category, and generalized flow paths in and around Rush Valley, Tooele County, Utah.

Summary

The water resources of Rush Valley were assessed during 2008–10 with an emphasis on refining the understanding of the groundwater-flow system and groundwater budget. Water resources are limited in Rush Valley with few perennial streams entering the valley from the mountains and no streams flowing through or exiting the valley. The limited surface-water resources generally are used for agriculture, leaving groundwater to supplement irrigation and as the predominant water source for most other uses. The principal source of groundwater in Rush Valley is from the unconsolidated basin fill, in which conditions are generally unconfined near the mountain front and confined in the lower-altitude parts of the valley. Productive aquifers are also present in bedrock. Although the majority of wells in the study area are completed within the basin fill, bedrock wells within the study area are increasingly being developed, especially in fractured bedrock beneath basin fill near Vernon, along the south flank of South Mountain, and in the Oquirrh Mountains.

Drillers' logs and geophysical (gravity) data were compiled and used to group geologic units into seven regionally important hydrogeologic units. Distinct basin-fill units include (1) the upper basin- fill aquifer unit (UBFAU), heterogeneous but laterally extensive unconsolidated Quaternary alluvial and lacustrine deposits; and (2) the lower basin-fill aquifer unit (LBFAU), a thick sequence of consolidated and semiconsolidated Tertiary-age lacustrine and alluvial deposits of the Salt Lake Formation. Regionally important bedrock units include (1) the upper carbonate aquifer unit (UCAU), interbedded Pennsylvanian to Permian-age limestone and sandstone collectively referred to as the Oquirrh Group; (2) an upper siliciclastic confining unit (USCU), upper Mississippian to lower Pennsylvanian-age shaly, siliciclastic, fine-grained sedimentary rocks comprised mostly of the Manning Canyon Shale; (3) the lower carbonate aquifer unit (LCAU), Middle Cambrian to Mississippian-age carbonates, sandstone and shale; (4) the noncarbonate confining unit (NCCU), a thick section of Precambrian to lower Cambrian-age sedimentary rocks consisting mostly of quartzite, shale, and diamictite; and (5) the volcanic unit (VU), grouped Tertiary-age intrusive and extrusive igneous rocks that are of limited extent within the study area. Most productive wells in Rush Valley are completed in the UBFAU and the UCAU.

Recharge occurs predominantly as direct infiltration of snowmelt and rainfall in the surrounding mountains (mountain recharge) or as infiltration of streamflow and unconsumed irrigation water at or near the mountain front (valley recharge). Average annual recharge to the Rush Valley groundwater basin was estimated to be 39,000 acre-ft. Groundwater generally moves from the higher altitude recharge areas toward two low-altitude discharge areas. A groundwater divide extends from the eastern edge of the Onaqui Mountains across the valley to near the mouth of Ophir Canyon, dividing the groundwater system such that most recharge occurring north of the divide discharges along the valley's central axis and in the vicinity of

Rush Lake, and most recharge occurring south of the divide discharges between Vernon and Faust. Most natural discharge occurs as evapotranspiration in these two areas. Groundwater discharge also occurs to mountain streams and springs, as well withdrawal, and although only a small fraction, as subsurface discharge to Tooele and Cedar Valleys. Average annual discharge from Rush Valley was estimated to be 43,000 acre-ft. The apparent recharge deficit of 4,000 acre-ft/year in the groundwater budget is most likely the result of uncertainty in the individual budget estimates.

Samples of groundwater collected from 25 sites were analyzed for major ions, nutrients, and selected trace metals to characterize general geochemical and water-quality patterns. Dissolved-solids concentrations in water ranged from 181 to 1,590 mg/L. More than half of the sites sampled during this study had dissolved-solids concentrations that exceeded the Environmental Protection Agency's secondary standard for drinking water of 500 mg/L. Seven of 25 sites sampled for arsenic contained concentrations that exceeded the EPA maximum contaminant level of 10 µg//L. Most waters with elevated arsenic concentrations are from wells located north of Vernon and in southeastern Rush Valley. The source of arsenic is likely from the mobilization of naturally occurring arsenic in aquifer sediments eroded from volcanic rocks in the surrounding mountains.

Water samples from the same 25 sites were also analyzed for a suite of environmental tracers that included the stable isotopes of oxygen, hydrogen, and carbon, dissolved noble gases, and radioactive isotopes of carbon (^{14}C) and hydrogen (tritium, ^{3}H) to investigate sources of recharge, groundwater-flow paths, ages, and traveltimes. Stable-isotope ratios of oxygen and deuterium and dissolved-gas recharge temperature data indicate that nearly all modern groundwater is meteoric and derived from the infiltration of mountain precipitation. These data support Basin Characterization Model estimates that show nearly all recharge occurring as infiltration of precipitation and snowmelt within the mountains surrounding Rush Valley. Concentrations of ^{3}H between 0.4 and 10 TU indicate the presence of modern (less than 60 years old) groundwater in samples from 7 of 25 sample sites. Apparent ^{3}H/^{3}He ages, calculated for six of these sites, ranged from 3 to 35 years. Adjusted minimum radiocarbon ages of premodern water samples range from 1,600 to 42,000 years with the ages of most (11 of 13) samples being more than 11,000 years. These data help to characterize three areas of more active groundwater movement that receive and circulate modern recharge on timescales of decades or less. They also indicate that large parts of the groundwater-flow system are much less active and receive little, if any, modern recharge. Much of the current groundwater withdrawal in Rush Valley occurs within or near the three areas of more active groundwater movement. These areas are replenished by modern recharge and, at the current rates of withdrawal, no long-term water-level declines have been observed. The surrounding, less active, parts of the groundwater-flow system appear to store potentially large, but unknown, quantities of water that are thousands to tens of thousands of years old. The age of groundwater in these areas indicates that they are not directly replenished by modern recharge. Therefore, monitoring of aquifer conditions as withdrawal continues or increases in the less active parts of the groundwater-flow system may be a valuable indicator of water-level declines that could represent irrecoverable groundwater-storage depletion.

References Cited

Aeschbach-Hertig, W., Peeters, F., Beyerle, U., and Kipfer, R., 1999, Interpretation of dissolved atmospheric noble gases in natural waters: Water Resources Research, v. 35, p. 2,779–2,792.

Aeschbach-Hertig, W., Peeters, F., Beyerle, U., and Kipfer, R., 2000, Paleotemperature reconstruction from noble gases in ground water taking into account equilibration with entrapped air: Nature, v. 405, p. 1,040–1,043.

Aiuppa, A., D'Alessandro, W., Federico, C., Palumbo, B., and Valenza, M., 2003, The aquatic geochemistry of arsenic in volcanic groundwater from southern Italy: Applied Geochemistry, v. 18, p. 1,283–1,296.

Aravena, R., Wassenaar, L.I., and Plummer, L.N., 1995, Estimation of C-14 groundwater ages in a methanogenic aquifer: Water Resources Research, v. 31, no. 9, p. 2,307–2,317.

Armstrong, R.L., 1968, Sevier orogenic belt in Nevada and Utah: Bulletin of the American Association of Petroleum Geologists, v. 79, p. 429–458.

Bagley, J.M., Jeppson, R.W., and Milligan, C.H., 1964, Water yields in Utah: Utah State University Agricultural Experiment Station Bulletin 439.

Ballentine, C.J., and Hall, C.M., 1999, Determining paleotemperature and other variables by using an error-weighted, nonlinear inversion of noble gas concentrations in water: Geochimica et Cosmochimica Acta, v. 63, p. 2,315–2,336.

Bankey, V., Grauch, V.J.S., and Kucks, R.P., 1998, Utah aeromagnetic and gravity maps and data—A web site for distribution of data: U.S. Geological Survey Open-File Report 98–761, 23 p.

Berger, D.L., Johnson, M.J., Tumbusch, M.L., and Mackay, J., 2001, Estimates of evapotranspiration from the Ruby Lake National Wildlife Refuge Area, Ruby Valley, northeastern Nevada, May 1999–October 2000: U.S. Geological Survey Water-Resources Investigations Report 01–4234, 38 p.

Burden, C.B., and others, 2009, Ground-water conditions in Utah, spring of 2009: Utah Department of Natural Resources, Division of Water Resources, Cooperative Investigations Report no. 50, 119 p.

Carpenter, Everett, 1913, Ground water in Boxelder and Tooele Counties, Utah: U.S. Geological Survey Water-Supply Paper 333, 90 p.

Casentini, B., Pettine, M., and Millero, F.J., 2010, Release of arsenic from volcanic rocks through interactions with inorganic anions and organic ligands: Aquatic Geochemistry, v. 16, p. 373–393.

Clark, D.L., Kirby, S.M., and Oviatt, C.G., 2009, Progress report geologic map of the Rush Valley 30' x 60' quadrangle, Tooele, Utah, and Salt Lake Counties, Utah (Year 1 of 3): Utah Geological Survey Open-File Report 555, 60 p., 1 pl., scale 1:62,500.

Clark, D.L., Kirby, S.M., and Oviatt, C.G., 2010, Progress report geologic map of the Rush Valley 30' x 60' quadrangle, Tooele, Utah, and Salt Lake Counties, Utah (Year 2 of 3): Utah Geological Survey Open-File Report 568, 83 p., 1 pl., scale 1:62,500.

Clark, D.W., and Appel, C.L., 1985, Ground-water resources of northern Utah Valley, Utah: Utah Department of Natural Resources Technical Publication no. 80, 115 p.

Clark, I., and Fritz, P., 1997, Environmental isotopes in hydrogeology: Boca Raton, Florida, CRC Press LLC, 328 p.

Constenius, K.N., 1996, Late Paleogene extensional collapse of the Cordilleran foreland fold and thrust belt: Geological Society of America Bulletin, v. 108, p. 20–39.

Constenius, K.N., Esser, R.P., and Layer, P.W., 2003, Extensional collapse of the Charleston-Nebo salient and its relationship to space-time variations in Cordilleran orogenic belt tectonism and continental stratigraphy, *in* Raynolds, R.G., and Flores, R.M., eds., Cenozoic systems of the Rocky Mountain region: Denver, Rocky Mountain Section: Society of Economic Paleontologists and Mineralogists, p. 303–353.

Cooper, D.J., Sanderson, J.S., Stannard, D.I., and Groeneveld, D.P., 2006, Effects of long-term water table drawdown on evapotranspiration and vegetation in an arid region phreatophyte community: Journal of Hydrology, v. 325, p. 21–34.

Cooper, H.H., and C.E. Jacob, 1946, A generalized graphical method for evaluating formation constants and summarizing well field history: American Geophysical Union Transactions, v. 27, p. 526–534

Coplen, T.B., 1994, Reporting of stable hydrogen, carbon, and oxygen isotopic abundances: Pure and Applied Chemistry, v. 66, no. 2, p. 273–276.

Craig, H., 1961a, Isotopic variations in meteoric waters: Science, v. 133, no. 3465, p. 1,702–1,703.

Craig, H., 1961b, Standard for reporting concentrations of deuterium and oxygen-18 in natural waters: Science, v. 133, no. 3467, p. 1,833–1,834.

Daly, C., Halbleib, M., Smith, J.I., Gibson, W.P., Doggett, M.K., Taylor, G.H., Curtis, J., and Pasteris, P.A., 2008, Physiographically-sensitive mapping of temperature and precipitation across the conterminous United States: International Journal of Climatology, v. 28, no. 15, p. 2,031–2,064. (Also available at http://dx.doi.org/10.1002/joc.1688.)

Dansgaard, W., 1964, Stable isotopes in precipitation: Talus, v. 16, p. 436–468.

DeCelles, P.G., and Coogan, J.C., 2006, Regional structure and kinematic history of the Sevier fold-and-thrust belt, central Utah: Geological Society of America Bulletin, v. 118, p. 841–864.

DeMeo, G.A., Smith, J.L., Damar, N.A., and Darnell, J., 2008, Quantifying ground-water and surface-water discharge from evapotranspiration processes in 12 hydrographic areas of the Colorado regional ground-water flow system, Nevada, Utah, and Arizona: U.S. Geological Survey Scientific Investigations Report 2008–5116, 22 p.

Domenico, P.A., and Schwartz, F.W., 1998, Physical and chemical hydrogeology: New York, John Wiley and Sons, 506 p.

Everitt, B.L. and Kaliser, B.N., 1980, Geology for the assessment of seismic risk in the Tooele and Rush Valleys, Tooele County, Utah: Utah Geological and Mineral Survey Special Study 51, 33 p., 7 pl., various scales.

Fairbanks, R.G., Mortlock, R.A., Chiu, T., Cao, L., Kaplan, A., Guilderson, T.P., Fairbanks, T.W., Bloom, A.L., Grootes, P.M., and Nadeau, M.J., 2005, Radiocarbon calibration curve spanning 0 to 50,000 years BP based on paired $^{230}Th/^{234}U/^{238}U$ and ^{14}C dates on pristine corals: Quaternary Science Reviews, v. 24, p. 1,781–1,796.

Feltis, R.D., 1967, Ground-water conditions in Cedar Valley, Utah County, Utah: Utah Department of Natural Resources Technical Publication no. 16, 34 p.

Fenneman, N.M., 1931, Physiography of the western United States: New York, McGraw-Hill, 534 p.

Flint, A.L., and Flint, L.E., 2007, Application of the basin characterization model to estimate in-place recharge and runoff potential in the Basin and Range carbonate-rock aquifer system, White Pine County, Nevada, and adjacent areas in Nevada and Utah: U.S. Geological Survey Scientific Investigations Report 2007–5099, 20 p.

Flint, A.L., and Flint, L.E., and Masbruch, M.D., 2011, Input, calibration, uncertainty, and limitations of the Basin Characterization Model, appendix 3 of Heilweil, V.M., and Brooks, L.E. (eds.), Conceptual model of the Great Basin carbonate and alluvial aquifer system: U.S. Geological Survey Scientific Investigations Report 2010–5193, 188 p.

Fontes, J.C., and Garnier, J.M., 1979, Determination of initial ^{14}C activity of the total dissolved carbon: A review of existing models and a new approach: Water Resources Research, v. 5, no. 2. p. 399–413.

Freeze, R.A., and Cherry, J.A., 1979, Groundwater: Englewood Cliffs, New Jersey, Prentice-Hall, 604 p.

Gardner, W.P., Susong D.D., Solomon, D.K., and Heasler, H., 2010, Snowmelt hydrograph interpretation: Revealing watershed scale hydrologic characteristics of the Yellowstone volcanic plateau: Journal of Hydrology, v. 383, p. 209–222.

Gates, J.S., 1963, Selected hydrologic data, Tooele Valley, Tooele County, Utah: U.S. Geological Open-File Report (duplicated as Utah Basic-Data Report No. 7), 23 p.

Gates, J.S., 1965, Re-evaluation of the ground-water resources of Tooele Valley, Utah: Utah State Engineer Technical Publication no. 12, 68 p.

Gates, J.S., 2007, Effects of climatic extremes on ground water in western Utah, 1930–2005: U.S. Geological Survey Scientific Investigations Report 2007–5045, 10 p.

Gilbert, G.K., 1890, Lake Bonneville: U.S. Geological Survey Monograph 1, 438 p.

Heilweil, V.M., and Brooks, L.E. (eds.), 2011, Conceptual model of the Great Basin carbonate and alluvial aquifer system: U.S. Geological Survey Scientific Investigations Report 2010–5193, 188 p.

Hely, A.J., Mower, R.W., and Harr, C.A., 1971, Water resources of Salt Lake County, Utah: Utah Department of Natural Resources Technical Publication no. 31, 224 p.

Hill, B.R., 1990, Ground water discharge to a headwater valley, northwestern Nevada, USA: Journal of Hydrology, v. 113, p. 265–283.

Hintze, L.F., Willis, G.C., Laes, D.Y.M., Sprinkel, D.A., and Brown, K.D., 2000, Digital geologic map of Utah: Utah Geological Survey Map 179DM.

Hood, J.W., Price, D., and Waddell, K.M., 1969, Hydrologic reconnaissance of Rush Valley, Tooele County, Utah: Utah Department of Natural Resources Technical Publication 23, 61 p.

Ingerson, R., and Pearson, F.J., 1964, Estimation of age and rate of motion of groundwater by the ^{14}C method, *in* Miyake, Y., and Koyama, T., eds., Recent research in the field of hydrosphere, atmosphere, and nuclear geochemistry: Tokyo, Marusen, 404 p.

International Atomic Energy Agency, 2007, Isotope hydrology section database: accessed February 2007, at http://www-naweb.iaea.org/napc/ih/.

Jeton, A.E., Watkins, S.A., Lopes, T.J., and Huntington, J., 2006, Evaluation of precipitation estimates from PRISM for the 1961–90 and 1971–2000 data sets, Nevada: U.S. Geological Survey Scientific Investigations Report 2005–5291, 26 p.

Kalin, R.M., 2000, Radiocarbon dating of groundwater systems, chap. 4 *of* Cook, P.G., and Herczeg A.L., eds., Environmental tracers in subsurface hydrology: Boston, Kluwer Academic Publishers, 529 p.

Karim, M., 2000, Arsenic in groundwater and health problems in Bangladesh: Water Resources, v. 34, no. 1, p. 304–310.

Keith, M.L., and Weber, J.N., 1964, Carbon and oxygen isotopic composition of selected limestones and fossils: Geochimica et Cosmochimica Acta, v. 28, p. 1,787–1,816.

Kendall, C., and Coplen, T.B., 2001, Distribution of oxygen-18 and deuterium in river waters across the United States: Hydrological Processes, v. 15, p. 1,363–1,393.

Kendall, C., and Caldwell, E.A., 1998, Fundamentals of isotope geochemistry, chap. 2 *of* Kendall, C., and McDonnell, J.J., eds., Isotope tracers in catchment hydrology: Amsterdam, Elsevier Science Publishers, 839 p.

Kipfer, R., Aeschbach-Hertig, W., Peeters, F., and Stute, M., 2002, Noble gases in lakes and ground waters, *in* Porcelli, D., Ballentine, C.J., and Wieler, R., eds., Reviews in mineralogy and geochemistry, v. 47, Noble gases in geochemistry and cosmochemistry: Mineralogical Society of America, Chantilly, Virginia, p. 615–700.

Kirby, S.M., 2010a, Interim geologic map of the Faust quadrangle, Tooele County, Utah: Utah Geological Survey Open File Report 573, 12 p., 2 pl., scale 1:24,000.

Kirby, S.M., 2010b, Interim geologic map of the of the Lofgreen quadrangle, Tooele County, Utah: Utah Geological Survey Open File Report 563, 17 p., 2 pl., scale 1:24,000.

Kirby, S.M., 2010c, Interim geologic map of the Saint John quadrangle, Tooele County, Utah: Utah Geological Survey Open File Report 572, 11 p., 2 pl., scale 1:24,000.

Kirby, S.M., 2010d, Interim geologic map of the Vernon quadrangle, Tooele County, Utah: Utah Geological Survey Open File Report 564, 18 p., 2 pl., scale 1:24,000.

Kirby, S.M., 2010e, Interim geologic map of the Vernon NE quadrangle, Tooele County, Utah: Utah Geological Survey Open File Report 562, 10 p., 2 pl., scale 1:24,000.

Lambert, P.M., and Stolp, B.J., 1999, Hydrology and simulation of the ground-water flow system in Tooele Valley, Utah: U.S. Geological Survey Water-Resources Investigations Report 99–4014, 60 p.

Levy, D.B., Schramke, J.A., Esposito, K.J., Erickson, T.A., and Moore, J.C., 1999, The shallow groundwater chemistry of arsenic, fluorine, and major elements: eastern Owens Lake, California: Applied Geochemistry, v. 14, p. 53–65.

Mahoney, J.R., 1953, Water resources of the Bonneville Basin, pt. 1, The water crop and its disposition: Utah Economic and Business Review, v. 13, no. 1–A.

Manning, A.H., and Solomon, D.K., 2003, Using noble gases to investigate mountain-front recharge: Journal of Hydrology, v. 275, p. 194–207.

Masbruch, M.D., Heilweil, V.M., Buto, S.G., Brooks, L.E., Susong, D.D., Flint, L.E., and Gardner, P.M., 2011, Groundwater budgets, chap. D *of* Heilweil, V.M., and Brooks, L.E. (eds.), Conceptual model of the Great Basin carbonate and alluvial aquifer system: U.S. Geological Survey Scientific Investigations Report 2010–5193, 188, p.

Mazor, E., and Bosch, A., 1992, Helium as a semi-quantitative tool for groundwater dating in the range of 10^4–10^8 years, *in* Isotopes of noble gases as tracers in environmental studies: Proceedings of a consultants meeting on isotopes of noble gases as tracers in environmental studies organized by the International Atomic Energy Agency, Vienna, May 29 to June 2, 1989: Vienna, International Atomic Energy Agency, p. 163–178.

Medina, R.L., 2005, 1:1,000,000-scale areas of evapotranspiration in the Great Basin: U.S. Geological Survey vector digital data, accessed March 2007, at http://water.usgs.gov/GIS/metadata/usgswrd/XML/ha694c_et1000gb_p.xml.

Michel, R.L., 1989, Tritium deposition over the continental United States, 1953–1983, *in* Atmospheric Deposition: International Association of Hydrological Sciences, p. 109–115.

Moreo, M.T., Laczniak, R.J., and Stannard, D.I., 2007, Evapotranspiration rate measurements of vegetation typical of ground-water discharge areas in the Basin and Range carbonate-rock aquifer system, Nevada and Utah, September 2005–August 2006: U.S. Geological Survey Scientific Investigations Report 2007–5078, 37 p.

Nichols, W.D., 2000, Determining ground-water evapotranspiration from phreatophyte shrubs and grasses as a function of plant cover or depth to ground water, Great Basin, Nevada and eastern California, chap. A of Nichols, W.D., ed., Regional ground-water evapotranspiration and ground-water budgets, Great Basin, Nevada: U.S. Geological Survey Professional Paper 1628, p. A1–A14.

North Wind, Inc., 2008, Final fall 2007 monitoring well sampling and analysis report, Deseret Chemical Depot, Utah: Prepared for U.S. Army Corps of Engineers, Sacramento District, Environmental Engineering Branch, 31 p.

Pan-American Center for Earth and Environmental Studies, 2010, GeoNet—United States Gravity Data Repository System: University of Texas El Paso, Pan-American Center for Earth and Environmental Studies, accessed February 2010, at http://paces.geo.utep.edu/gdrp.

Parkhurst, D.L., and Plummer, L.N., 1983, Geochemical models, chap. 9 of Alley, W.M., eds., Regional groundwater quality: New York, Van Nostrand Reinhold, 634 p.

Pearson, F.J., and White, D.E., 1967, Carbon-14 ages and flow rates in Carrizo Sand, Atascosa County, Texas: Water Resources Research, v. 3, no. 1, p. 251–261.

Planer-Friedrich, B., Armienta, M.A., and Merkel, B.J., 2001, Origin of arsenic in the groundwater of the Rioverde basin, Mexico: Environmental Geology, v. 40, p. 1,290–1,298.

Plummer, L.N., Busby, J.F., Lee, R.W., and Hanshaw, B.B., 1990, Geochemical modeling of the Madison aquifer in parts of Montana, Wyoming, and South Dakota: Water Resources Research v. 26, no. 9, p. 1,981–2,014.

Plummer, L.N., Prestemon, E.C., and Parkhurst, D.L., 1994, An interactive code (NETPATH) for modeling net geochemical reactions along a flow path, version 2.0: U.S. Geological Survey Water-Resources Investigations Report 94–4169, 130 p.

Plummer, L.N., and Sprinkle, C.L., 2001, Radiocarbon dating of dissolved inorganic carbon in groundwater from confined parts of the Upper Floridan aquifer, Florida, USA: Hydrogeology Journal, v. 9, p. 127–150.

Razem, A.C., and Steiger, J.I., 1981, Ground-water conditions in Tooele Valley, Utah, 1976–78: Utah Department of Natural Resources Technical Publication no. 69, 95 p.

Reiner, S.R., Laczniak, R.J., DeMeo, G.A., Smith, J.L., Elliott, P.E., Nylund, W.E., and Fridrich, C.J., 2002, Ground-water discharge determined from measurements of evapotranspiration, other available hydrologic components, and shallow water-level changes, Oasis Valley, Nye County, Nevada: U.S. Geological Survey Water-Resources Investigations Report 01–4239, 65 p.

Ryu, J., Gao, S., Dahlgren, R., and Zierenberg, R., 2002, Arsenic distribution, speciation and solubility in shallow groundwater of Owens Dry Lake, California: Geochimica et Cosmochimica Acta, v. 66, no. 17, p. 2,981–2,994.

Sheldon, A., 2002, Diffusion of radiogenic helium in shallow ground water—Implications for crustal degassing: Salt Lake City, University of Utah, Ph.D. dissertation, 185 p.

Smedley, P.L., and Kinniburgh, D.G., 2002, A review of the source, behaviour and distribution of arsenic in natural waters: Applied Geochemistry, v. 17, p. 517–568.

Smedley, P.L., Nicolli, H.B., Macdonald, D.M.J., Barros, A.J., and Tullio, J.O., 2002, Hydrogeochemistry of arsenic in groundwater from La Pampa, Argentina: Applied Geochemistry, v. 17, p. 259–284.

Smith, G.I., Friedman, I., Veronda, G., and Johnson, C.A., 2002, Stable isotope compositions of waters in the Great Basin, United States 3. Comparison of groundwaters with modern precipitation: Journal of Geophysical Research, v. 107, p. ACL16.1–ACL16.15.

Snyder, C.T., 1963, Hydrology of stock-water development on the public domain of western Utah: U.S. Geological Survey Water-Supply Paper 1475–N, p. 487–536.

Solomon, D.K., 2000, ^{4}He in groundwater, chap. 14 of Cook, P.G., and Herczeg A.L., eds., Environmental tracers in subsurface hydrology: Boston, Kluwer Academic Publishers, 529 p.

Solomon, D.K., and Cook, P.G., 2000, ^{3}H and ^{3}He, chap. 13 of Cook, P.G., and Herczeg A.L., eds., Environmental tracers in subsurface hydrology: Boston, Kluwer Academic Publishers, p. 397–424.

Stolp, B.J., 1994, Hydrology and potential for ground-water development in southeastern Tooele Valley and adjacent areas in the Oquirrh Mountains, Tooele County, Utah: Utah Department of Natural Resources Technical Publication no. 107, 67 p.

Stolp, B.J., and Brooks, L.E., 2009, Hydrology and simulation of ground-water flow in the Tooele Valley ground-water basin, Tooele County, Utah: U.S. Geological Survey Scientific Investigations Report 2009–5154, 86 p., available online at http://pubs.usgs.gov/sir/2009/5154/.

Stute, M., and Schlosser, P., 2000, Atmospheric noble gases, chap. 11 of Cook, P.G., and Herczeg, A.L., eds., Environmental tracers in subsurface hydrology: Boston, Kluwer Academic Publishers, p. 349–377.

Susong, D.D., 1995, Water budget and simulation of one-dimensional unsaturated flow in a flood- and sprinkler-irrigated field near Milford, Utah: Utah Department of Natural Resources Technical Publication no. 109, 32 p.

Sweetkind, D.S., Cederberg, J.R., Masbruch, M.D., and Buto, S.G., 2011, Hydrogeologic framework, chap. B *of* Heilweil, V.M., and Brooks, L.E. eds., Conceptual model of the Great Basin carbonate and alluvial aquifer system: U.S. Geological Survey Scientific Investigations Report 2010–5193, 188 p.

Tague, C., and Grant, G. E., 2004, A geological framework for interpreting the low-flow regimes of Cascade streams, Willamette River Basin, Oregon: Water Resources Research, v. 40, W04303. (Also available at http://dx.doi.org/10.1029/2003WR002629.)

Tamers, M.A., 1975, Validity of radiocarbon dates on groundwater: Geophysical Surveys, v. 2, p. 217–239.

Thomas, H.E., 1946, Ground water in Tooele County, Tooele County, Utah: Utah State Engineer Technical Publication no. 4, in Utah State Engineer 25th Biennial Report, p. 91–238.

U.S. Census Bureau, 2000, Census 2000 summary file (SF 1) 100-percent data, accessed January 25, 2009, at http://factfinder.census.gov/servlet/GCTTable?_bm=y&-geo_id=04000US49&-_box_head_nbr=GCT-PH1&-ds_name=DEC_2000_SF1_U&-format=ST-7&-mt_name=DEC_2000_SF1_U_GCTPH1_ST2.

U.S. Department of Agriculture, 2006, National Agricultural Imagery Program (NAIP), Compressed County mosaics (CCM) for Utah, accessed January 20, 2011, at http://datagateway.nrcs.usda.gov/GDGOrder.aspx.

U.S. Environmental Protection Agency, 2007, Southwest Regional Gap Analysis Program (SWReGAP) for Nevada, Utah, New Mexico, Colorado, and Arizona— Project status: U.S. Environmental Protection Agency, Office of Research and Development, Las Vegas, Nevada, accessed May 10, 2009, at http://www.epa.gov/esd/land-sci/gap-status.htm.

U.S. Environmental Protection Agency, 2009, Drinking water contaminants: accessed September 2009, at http://www.epa.gov/safewater/contaminants/index.html#mcls.

U.S. Geological Survey, 1991, National water summary, 1988–89: U.S. Geological Survey Water-Supply Paper 2375, 591 p.

Utah Department of Natural Resources, 2004, Downloadable GIS land use data: accessed April 2006, at http://www.water.utah.gov/planning/landuse/gisdata.htm.

Utah Department of Natural Resources, 2010, Utah Division of Water Rights well-drilling database: accessed January 2010, at http://www.waterrights.utah.gov/cgi-bin/wellview.exe.

Utah State University, 1994, Consumptive use of irrigated crops in Utah: Utah Agricultural Experiment Station Research Report 145, Logan Utah State University, 361 p.

Watershed Boundary Dataset, Tooele County, Utah: accessed March 2008, at http://datagateway.nrcs.usda.gov.

Welch, A.H., Bright, D.J., and Knochenmus, L.A., eds., 2007, Water resources of the Basin and Range carbonate-rock aquifer system, White Pine County, Nevada, and adjacent areas in Nevada and Utah: U.S. Geological Survey Scientific Investigations Report 2007–5261, 96 p.

Wilde, F.D., and Radtke, D.B., 1998, Field measurements: U.S. Geological Survey Techniques of Water-Resources Investigations, book 9, chap. A6 [variously paged].

Wilkowske, C.D., Kenney, T.A., and Wright, S.J., 2008, Methods for estimating monthly and annual streamflow statistics at ungaged sites in Utah: U.S. Geological Survey Scientific Investigations Report 2008–5230, 63 p., available online at http://pubs.usgs.gov/sir/2008/5230/.

Appendix 1. Data Tables

Table A1–1. Selected physical attributes of wells and water levels in wells where synoptic water-level measurements were made during the fall of 2008 or spring of 2009 in and around Rush Valley, Tooele County, Utah.

[Location: See "numbering system" at beginning of report for an explanation of the numbering system used for hydrologic-data sites in Utah Hydrogeologic unit and aquifer condition: UBFAU, upper basin-fill aquifer unit; LBFAU, lower basin-fill aquifer unit; UCAU, upper carbonate aquifer unit; LCAU, lower carbonate aquifer unit; C, confined; U, unconfined; ?, uncertain; —, no information; OB, open bottom]

Location	Depth of well, in feet	Depth to top and bottom of openings, in feet	Altitude of land surface, in feet	Hydrogeologic unit that well is completed in and aquifer condition	Date measured	Water level, in feet below land surface	Water-level altitude, in feet	Date measured	Water level, in feet below land surface	Water-level altitude, in feet	Water-level change, in feet[1]
(C-4-4)32add-1	620	420–620	5,760	UCAU, ?	10/6/2008	290.07	5,469.9	—	—	—	—
(C-4-4)32bcd-1	340	280–340	5,520	UCAU, ?	10/6/2008	165.94	5,354.1	3/16/2009	178.82	5,341.2	-12.9
(C-4- 4)33cbd-2	200	190-200	5,755	LCAU, ?	10/16/2008	148.53	5,606.5	3/16/2009	166.25	5,588.8	-17.7
(C-4-5)13bad-1	338	—	4,940	UBFAU, U	10/6/2008	94.55	4,845.5	3/16/2009	95.08	4,844.9	-0.6
(C-4-5)27cdb-1	410	127–410	5,000	UCAU, ?	10/7/2008	186.97	4,813.0	3/16/2009	187.09	4,812.9	-0.1
(C-4-5)29bdc-2	700	500–700	5,210	UCAU, ?	10/6/2009	444.05	4,766.0	3/16/2009	444.24	4,765.8	-0.2
(C-4-5)30aac-1	704	694–704	5,247	UCAU, ?	10/7/2008	482.66	4,764.3	3/16/2009	482.11	4,764.9	0.6
(C-4-6)1dbb-1	780	718–768	5,033	UBFAU, C	10/16/2008	626.74	4,406.7	3/31/2009	627.61	4,405.8	-0.9
(C-4-6)15cac-1	260	200–240	5,620	UCAU, ?	10/7/2008	188.69	5,431.3	3/16/2009	188.95	5,431.1	-0.2
(C-4-6)23dbd-1	502	440–502	5,528	UBFAU, U	10/9/2008	329.54	5,198.5	3/16/2009	330.64	5,197.4	-1.1
(C-4-6)23dda-1	450	OB	5,495	UBFAU, U	10/7/2008	298.94	5,196.1	3/16/2009	299.77	5,195.2	-0.9
(C-4-6)24cad-1	510	450–490	5,395	UCAU, ?	10/6/2008	213.88	5,181.1	3/16/2009	213.70	5,181.3	0.2
(C-4-6)28aad-1	520	460–520	6,010	UBFAU, U	10/7/2008	420.93	5,589.1	3/16/2009	421.55	5,588.5	-0.6
(C-5-4)28cdb-1	90	67–88	5,640	LCAU, ?	10/16/2008	67.60	5,572.4	3/20/2009	69.69	5,570.3	-2.1
(C-5-5)5adb-1	209	51–168	4,985	UBFAU, C	11/3/2008	3.78	4,981.2	3/17/2009	2.24	4,982.8	1.6
(C-5-5)5bdb-1	200	32–200	5,010	UBFAU, C	10/14/2008	37.06	4,972.9	3/17/2009	34.50	4,975.5	2.6
(C-5-5)11bba-1	238	211–236	5,048	UBFAU, U	10/9/2008	76.60	4,971.4	3/24/2009	76.83	4,971.2	-0.2
(C-5-5)15add-1	100	10–?	5,030	UBFAU, C	10/15/2008	3.46	5,026.5	3/17/2009	3.01	5,027.0	0.5
(C-5-5)17aad-1	20	—	4,995	UBFAU, C	10/8/2008	9.95	4,985.1	3/17/2009	7.75	4,987.3	2.2
(C-5-5)19cab-1	145	105–145	5,075	UBFAU, C	10/28/2008	12.87	5,062.1	3/20/2009	11.28	5,063.7	1.6
(C-5-5)20aba-1	60	—	5,010	UBFAU, C	10/9/2008	4.31	5,005.7	3/17/2009	7.11	5,002.9	-2.8
(C-5-5)20abb-1	200	38–200	5,015	UBFAU, C	11/3/2008	4.70	5,010.3	3/17/2009	23.75	4,991.3	-19.0
(C-5-5)20acc-2	245	—	5,016	UBFAU, C	11/3/2008	5.50	5,010.5	3/17/2009	0.70	5,015.3	4.8
(C-5-5)20bcc-1	21	—	5,020	UBFAU, C	10/7/2008	4.52	5,015.5	3/17/2009	1.45	5,018.6	3.1
(C-5-5)20daa-1	—	—	5,015	UBFAU, C	10/8/2008	17.79	4,997.2	3/18/2009	17.51	4,997.5	0.3
(C-5-5)20daa-2	35	24–34	5,016	UBFAU, C	10/8/2008	15.47	5,000.5	3/18/2009	14.66	5,001.3	0.8
(C-5-5)29daa-1	33	23–33	5,025	UBFAU, C	10/8/2008	19.31	5,005.7	3/17/2009	19.20	5,005.8	0.1
(C-5-5)30bda-2	21	16–21	5,052	UBFAU, C	10/7/2008	20.70	5,031.3	3/17/2009	17.18	5,034.8	3.5
(C-5-5)30caa-1	66	34–64	5,064	UBFAU, C	10/8/2008	30.10	5,033.9	3/20/2009	24.36	5,039.6	5.7
(C-5-5)30cac-1	572	115–513	5,075	UBFAU, C	10/8/2008	33.55	5,041.5	3/18/2009	29.02	5,046.0	4.5
(C-5-5)30dac-1	79	39–79	5,036	UBFAU, C	10/28/2008	16.01	5,020.0	3/24/2009	13.12	5,022.9	2.9
(C-5-5)30dbc-1	88	28–88	5,064	UBFAU, ?	10/7/2008	32.85	5,031.2	3/24/2009	30.13	5,033.9	2.7
(C-5-5)30dbd-1	100	80–100	5,050	UBFAU, C	10/8/2008	27.28	5,022.7	3/20/2009	23.50	5,026.5	3.8
(C-5-5)30ddb-1	105	65–105	5,040	UBFAU, C	10/8/2008	19.96	5,020.0	3/20/2009	18.65	5,021.4	1.4
(C-5-5)31acc-1	215	110–215	5,100	UBFAU, C	10/24/2008	24.20	5,075.8	—	—	—	—
(C-5-5)32cac-2	145	77–?	5,048	UBFAU, C	10/7/2008	27.78	5,020.2	3/17/2009	23.02	5,025.0	4.8
(C-5-5)33bcc-1	30	26–30	5,025	UBFAU, C	10/8/2008	12.95	5,012.1	3/17/2009	12.51	5,012.5	0.4
(C-5-6)12aba-1	—	—	5,250	UBFAU, U	10/7/2008	58.41	5,191.6	3/17/2009	58.77	5,191.2	-0.4

Table A1–1.   Selected physical attributes of wells and water levels in wells where synoptic water-level measurements were made during the fall of 2008 or spring of 2009 in and around Rush Valley, Tooele County, Utah.—Continued

[Location: See "numbering system" at beginning of report for an explanation of the numbering system used for hydrologic-data sites in Utah Hydrogeologic unit and aquifer condition: UBFAU, upper basin-fill aquifer unit; LBFAU, lower basin-fill aquifer unit; UCAU, upper carbonate aquifer unit; LCAU, lower carbonate aquifer unit; C, confined; U, unconfined; ?, uncertain; —, no information; OB, open bottom]

Location	Depth of well, in feet	Depth to top and bottom of openings, in feet	Altitude of land surface, in feet	Hydrogeologic unit that well is completed in and aquifer condition	Date measured	Water level, in feet below land surface	Water-level altitude, in feet	Date measured	Water level, in feet below land surface	Water-level altitude, in feet	Water-level change, in feet[1]
(C-5-6)25aaa-1	108	104–108	5,125	UBFAU, C	10/14/2008	16.12	5,108.9	3/17/2009	10.74	5,114.3	5.4
(C-5-6)25caa-1	247	124–134, 242–247	5,244	UBFAU, U	10/7/2008	60.30	5,183.7	3/24/2009	60.05	5,184.0	0 3
(C-5-6)35dab-1	126	86–126	5,363	UBFAU, C	10/22/2008	21.81	5,341.2	3/24/2009	22.73	5,340.3	-0 9
(C-6-2)29cac-2	220	—	4,889	UBFAU, C	10/15/2008	0.39	4,888.3	—	—	—	—
(C-6-4)4acc-1	1010	500–1,000	5,598	UBFAU, U	10/28/2008	484.47	5,113.5	3/20/2009	484.19	5,113.8	0 3
(C-6-4)31ddc-1	100	30–50	5,045	UBFAU, C	10/27/2008	33.34	5,011.7	3/18/2009	30.85	5,014.2	2 5
(C-6-4)35bac-1	365	—	5,135	UBFAU, U	10/9/2008	141.25	4,993.8	3/19/2009	141.42	4,993.6	-0 2
(C-6-5) 5bab-1	70	35–45	5,055	UBFAU, C	10/15/2008	18.28	5,036.7	3/20/2009	17.46	5,037.5	0.8
(C-6-5)6aab-1	120	15–120	5,080	UBFAU, C	10/29/2008	9.95	5,070.1	3/20/2009	9.45	5,070.6	0 5
(C-6-5)6ada-1	180	132–140	5,072	UBFAU, C	10/14/2008	15.20	5,056.8	3/20/2009	14.34	5,057.7	0 9
(C-6-5)7aab-1	207	190–207	5,107	UBFAU, C	10/29/2008	26.21	5,080.8	3/20/2009	25.73	5,081.3	0 5
(C-6-5)9aca-1	31	21–31	5,025	UBFAU, C	10/9/2008	9.30	5,015.7	3/17/2009	8.23	5,016.8	1 1
(C-6-5)16dbd-1	38	28–38	5,035	UBFAU, C	10/14/2008	10.10	5,024.9	3/17/2009	9.71	5,025.3	0.4
(C-6-5)27ccb-1	30	25–30	5,065	UBFAU, C	10/15/2008	23.64	5,041.4	3/18/2009	22.64	5,042.4	1.0
(C-6-5)28add-1	120	100–120	5,064	UBFAU, C	10/30/2008	17.78	5,046.2	3/18/2009	17.14	5,046.9	0.7
(C-6-5)34ddd-1	140	100–140	5,092	UBFAU, C	10/27/2008	29.85	5,062.2	—	—	—	—
(C-6-6)1aaa-1	—	—	5,184	—	10/29/2008	47.13	5,136.9	3/20/2009	44.93	5,139.1	2 2
(C-6-6)1bcc-1	140	110–140	5,345	UBFAU, ?	10/14/2008	28.36	5,316.6	3/20/2009	27.97	5,317.0	0.4
(C-6-6)11ccc-1	128	100–128	5,476	UBFAU, U	10/20/2008	79.60	5,396.4	3/20/2009	71.63	5,404.4	8.0
(C-6-6)12bcb-1	69	45–70	5,310	UBFAU, U	10/14/2008	41.77	5,268.2	3/20/2009	40.71	5,269.3	1 1
(C-6-6)12bcc-1	105	—	5,310	UBFAU, U	10/14/2008	37.99	5,272.0	3/20/2009	37.44	5,272.6	0.6
(C-6-6)24dbb-1	120	30–120	5,290	UBFAU, U	10/20/2008	67.87	5,222.1	3/20/2009	67.85	5,222.2	0 1
(C-6-7)3cba-1	480	400–480	5,250	—	10/9/2008	310.07	4,939.9	—	—	—	—
(C-6-7)3cdb-1	256	212–254	5,230	—	10/16/2008	181.60	5,048.4	—	—	—	—
(C-6-7)16cbb-1	300	150–295	4,871	—	10/16/2008	121.08	4,749.9	3/17/2009	123.07	4,747.9	-2.0
(C-7-2)23bcc-2	275	235–275	4,835	—	10/15/2008	113.92	4,721.1	3/18/2009	113.88	4,721.1	0.0
(C-7-2)29dbc-1	198	—	4,860	—	10/9/2008	165.33	4,694.7	3/18/2009	165.61	4,694.4	-0 3
(C-7-2)31abc-1	—	—	4,880	—	3/4/2009	188.63	4,691.4	—	—	—	—
(C-7-2)33aaa-1	400	OB	4,852	—	10/14/2008	170.33	4,681.7	3/18/2009	170.28	4,681.7	0.0
(C-7-3)30acc-1	300	140–?	5,080	UBFAU, U	10/9/2008	129.91	4,950.1	3/19/2009	130.90	4,949.1	-1.0
(C-7-4)35cdd-1	325	—	5,135	UBFAU, U	10/9/2008	184.38	4,950.6	3/19/2009	184.04	4,951.0	0.4
(C-7-5)16aab-1	120	100–120	5,139	UBFAU, C	10/16/2008	16.51	5,122.5	3/18/2009	14.23	5,124.8	2 3
(C-7-5)16ada-1	81	64–71	5,130	UBFAU, C	10/29/2008	5.65	5,124.4	3/20/2009	3.49	5,126.5	2 1
(C-7-5)28ccc-1	315	260–300	5,224	UBFAU, C	11/3/2008	4.52	5,219.5	3/20/2009	3.70	5,220.3	0.8
(C-7-5)29dca-1	250	142–250	5,263	UBFAU, C	10/16/2008	33.25	5,229.8	3/18/2009	32.34	5,230.7	0 9
(C-7-5)29ddb-1	262	257–262	5,250	UBFAU, C	10/15/2008	21.05	5,229.0	3/18/2009	20.54	5,229.5	0 5
(C-7-5)32bdd-1	360	—	5,272	LBFAU, C	10/9/2008	12.09	5,259.9	3/20/2009	11.01	5,261.0	1 1
(C-8-2)31aac-1	365	—	5,010	—	10/14/2008	341.52	4,668.5	—	—	—	—
(C-8-3)3cbd-1	1,390	430–498	5,312	LCAU, ?	10/15/2008	342.67	4,969.3	3/19/2009	342.14	4,969.9	0.6
(C-8-4)5cbd-1	—	—	5,252	—	10/28/2008	81.56	5,170.4	3/19/2009	81.41	5,170.6	0 2
(C-8-4)36aab-1	1,002	800–1,000	5,337	LBFAU, U	10/28/2008	314.88	5,022.1	3/18/2009	319.01	5,018.0	-4 1

Table A1–1. Selected physical attributes of wells and water levels in wells where synoptic water-level measurements were made during the fall of 2008 or spring of 2009 in and around Rush Valley, Tooele County, Utah.—Continued

[Location: See "numbering system" at beginning of report for an explanation of the numbering system used for hydrologic-data sites in Utah Hydrogeologic unit and aquifer condition: UBFAU, upper basin-fill aquifer unit; LBFAU, lower basin-fill aquifer unit; UCAU, upper carbonate aquifer unit; LCAU, lower carbonate aquifer unit; C, confined; U, unconfined; ?, uncertain; —, no information; OB, open bottom]

Location	Depth of well, in feet	Depth to top and bottom of openings, in feet	Altitude of land surface, in feet	Hydrogeologic unit that well is completed in and aquifer condition	Date measured	Water level, in feet below land surface	Water-level altitude, in feet	Date measured	Water level, in feet below land surface	Water-level altitude, in feet	Water-level change, in feet[1]
(C-8-5)6ccd-1	730	132–?	5,300	UBFAU and UCAU, C	11/6/2008	53.24	5,246.8	3/20/2009	47.21	5,252.8	6.0
(C-8-5)7ddd-2	547	—	5,360	UBFAU and UCAU, C	11/7/2008	119.00	5,241.0	3/24/2009	110.29	5,249.7	8.7
(C-8-5)17cad-1	210	113–210	5,433	UBFAU, C	10/28/2008	62.73	5,370.3	3/20/2009	62.76	5,370.2	0.1
(C-8-5)17ccc-1	800	634–800	5,413	UCAU, C	11/7/2008	171.93	5,241.1	3/24/2009	165.64	5,247.4	6.3
(C-8-5)19ddb-1	126	106–126	5,455	UBFAU, C	10/20/2008	2.41	5,452.6	3/19/2009	2.94	5,452.1	-0.5
(C-8-5)20bbd-1	282	219–280	5,425	UBFAU, C	11/7/2008	17.04	5,408.0	3/24/2009	17.31	5,407.7	-0.3
(C-8-5)20bca-1	284	32–280	5,433	UBFAU, C	10/15/2008	13.80	5,419.2	3/24/2009	13.82	5,419.2	0.0
(C-8-5)20cdd-1	212	195–212	5,479	UBFAU, C	10/9/2008	13.00	5,466.0	3/20/2009	13.63	5,465.4	-0.6
(C-8-5)20dcc-1	275	137–270	5,508	UBFAU, C	11/7/2008	67.61	5,440.4	3/24/2009	67.60	5,440.4	0.0
(C-8-5)30cac-1	53	—	5,508	UBFAU, C	11/7/2008	9.78	5,498.2	3/20/2009	8.65	5,499.4	1.2
(C-8-5)31ccd-5	60	—	5,576	UBFAU, C	11/7/2008	19.68	5,556.3	3/19/2009	15.60	5,560.4	4.1
(C-8-5)31cdc-1	150	80–150	5,584	UBFAU, C	10/16/2008	24.40	5,559.6	3/19/2009	22.24	5,561.8	2.2
(C-8-6)12aca-1	151	46–146	5,330	UBFAU, C	10/31/2008	20.62	5,309.4	3/19/2009	16.17	5,313.8	4.4
(C-8-6)12acb-1	220	160–220	5,335	UBFAU, C	10/31/2008	16.70	5,318.3	3/19/2009	13.73	5,321.3	3.0
(C-8-6)21cab-1	343	303–340	5,632	UBFAU, U	10/31/2008	254.68	5,377.3	3/19/2009	254.35	5,377.7	0.4
(C-8-6)23ccd-1	—	—	5,448	UBFAU, C	10/15/2008	19.80	5,428.2	3/19/2009	26.71	5,421.3	-6.9
(C-8-6)25bab-1	253	—	5,440	UBFAU, C	11/7/2008	4.43	5,435.6	3/19/2009	8.31	5,431.7	-3.9
(C-8-6)26aaa-1	224	—	5,426	UBFAU, C	10/15/2008	16.18	5,409.8	3/19/2009	13.90	5,412.1	2.3
(C-8-6)34acc-1	180	30–130	5,606	UBFAU, U	10/30/2008	61.47	5,544.5	3/19/2009	61.61	5,544.4	-0.1
(C-8-6)35bca-2	150	—	5,518	UBFAU, C	10/16/2008	13.31	5,504.7	3/19/2009	9.49	5,508.5	3.8
(C-8-6)36cdd-1	333	329–333	5,586	UBFAU, C	10/30/2008	65.43	5,520.6	3/19/2009	64.44	5,521.6	1.0
(C-9-4)15bbc-1	120	80–120	5,638	UBFAU, U	10/30/2008	37.94	5,600.1	3/20/2009	37.82	5,600.2	0.1
(C-9-5)5bbc-1	600	Uncased below 493	5,595	UBFAU, C	10/9/2008	14.41	5,580.6	3/20/2009	8.07	5,586.9	6.3
(C-9-5)6aab-1	—	—	5,585	UBFAU, C	10/9/2008	37.78	5,547.2	3/20/2009	30.67	5,554.3	7.1
(C-9-5)6aab-2	170	—	5,597	UBFAU, C	10/9/2008	37.85	5,559.2	3/20/2009	31.19	5,565.8	6.6
(C-9-6)1bdc-1	75	15–?	5,625	UBFAU, C	10/16/2008	7.10	5,617.9	3/19/2009	5.84	5,619.2	1.3
(C-10-4)14aab-1	223	180–223	6,073	UBFAU, U	10/28/2008	129.96	5,943.0	—	—	—	—
S-93-92[2]	—	136–152	5,077	UBFAU, U	Oct-08	—	5,011.2	—	—	—	—
S-45-90[2]	—	21–31	5,053	UBFAU, C	Oct-08	—	5,036.3	—	—	—	—
S-63-90[2]	—	84–104	5,122	UBFAU, U	Oct-08	—	5,032.7	—	—	—	—
S-50-90[2]	—	57–67	5,157	UBFAU, U	Oct-08	—	5,093.7	—	—	—	—
S-56-90[2]	—	39–49	5,056	UBFAU, C	Oct-08	—	5,033.6	—	—	—	—
S-106-93[2]	—	41–51	5,044	UBFAU, C	Oct-08	—	5,031.0	—	—	—	—
S-49-90[2]	—	99–109	5,144	UBFAU, U	Oct-08	—	5,060.9	—	—	—	—
S-8[2]	—	65–85	5,194	UBFAU, U	Oct-08	—	5,111.3	—	—	—	—
S-95-92[2]	—	103–128	5,050	UBFAU, C	Oct-08	—	5,019.9	—	—	—	—
S-41-90[2]	—	287–307	5,382	UBFAU, U	Oct-08	—	5,095.8	—	—	—	—
S-32-90[2]	—	215–235	5,330	UBFAU, U	Oct-08	—	5,113.7	—	—	—	—
S-12-88[2]	—	34–38	5,056	UBFAU, C	Oct-08	—	5,037.9	—	—	—	—
S-14[2]	—	12–32	5,042	UBFAU, C	Oct-08	—	5,029.2	—	—	—	—
S-BR-1[2]	—	109–149	5,233	UBFAU, U	Oct-08	—	5,110.2	—	—	—	—

Table A1–2.   Selected physical attributes of groundwater sites sampled during the summer of 2008 or 2009, and of sites with monthly and long-term water-level time-series data in and around Rush Valley, Tooele County, Utah.

[Location: See "numbering system" at beginning of report for an explanation of the numbering system used for hydrologic-data sites in Utah Hydrogeologic unit and aquifer condition: UBFAU, upper basin-fill aquifer unit; LBFAU, lower basin-fill aquifer unit; UCAU, upper carbonate aquifer unit; LCAU, lower carbonate aquifer unit; C, confined; U, unconfined Type of data: M, monthly water levels; GC, geochemical Sample identifier: relates physical site information to groundwater geochemical data presented in tables 8–10 and figures 15–23 See figure 15 for the location of sites sampled as part of this study —, no information; ?, uncertain ; >, greater than]

Location	Site ID	Depth of well, in feet	Depth to top and bottom of openings, in feet	Altitude of land surface, in feet	Hydrogeologic unit that well is completed in and aquifer condition	Sub area	Type of data
(C-4-5)8cbb-1	25	715	uncased below 640	5,080	UCAU, ?	Tooele Valley	GC
(C-4-5)13bad-1	24	338	—	4,940	UBFAU, U	Tooele Valley	M, GC
(C-4-5)27cdb-1	—	410	127–410	5,000	UCAU, ?	Northern Rush Valley	M
(C-4-5)29bdc-2	—	700	500–700	5,210	UCAU, ?	Northern Rush Valley	M
(C-4-5)30aac-2	19	710	545–705	5,245	UCAU, ?	Northern Rush Valley	GC
(C-4-4)32add-1	20	620	420–620	5,760	UCAU, ?	Northern Rush Valley	GC
(C-5-5) 3bcc-1	23	300	282–300	4,965	UBFAU, C	Northern Rush Valley	GC
(C-5-5) 8dab-1	18	>100	—	4,985	UBFAU, C	Northern Rush Valley	GC
(C-5-5)11bba-1	22	238	211-236	5,048	UBFAU, U	Northern Rush Valley	GC
(C-5-5)15add-2	21	53	53–open end	5,030	UBFAU, C	Northern Rush Valley	GC
(C-5-5)32cac-2	—	145	77–?	5,048	UBFAU, C	Northern Rush Valley	M
(C-5-5)32dbb-2	17	70	40–70	5,053	UBFAU, C	Northern Rush Valley	GC
(C-5-6)12aba-1	16	192	—	5,250	UBFAU, U	Northern Rush Valley	GC
(C-5-6)32bba-S1	14	—	—	5,900	LCAU, C	Northern Rush Valley	GC
(C-6-4)35bac-1	—	365	—	5,135	UBFAU, U	Southeastern Rush Valley	M
(C-6-6)11ccc-1	15	128	100–128	5,476	UBFAU, U	Northern Rush Valley	GC
(C-7-2)29dbc-1	13	198	—	4,860	—	Cedar Valley	M, GC
(C-7-3)30acc-1	11	300	140–?	5,080	UBFAU, U	Southeastern Rush Valley	M, GC
(C-7-5)28ccc-1	6	315	260–300	5,224	UBFAU, C	Vernon area	GC
(C-7-5)32bdd-1	—	360	—	5,272	LBFAU, C	Vernon area	M
(C-8-3)3cbd-1	12	1,390	430–498	5,312	LCAU, ?	Southeastern Rush Valley	GC
(C-8-4)22aad-1	9	298	—	5,290	UBFAU, U	Southeastern Rush Valley	GC
(C-8-4)36aab-1	10	1,002	800–1,000	5,337	LBFAU, U	Southeastern Rush Valley	GC
(C-8-5)6ccd-1	5	730	132–?	5,300	UBFAU and UCAU, C	Vernon area	GC
(C-8-5)17ccc-1	4	800	634–800	5,413	UCAU, C	Vernon area	GC
(C-8-5)20cdd-1	—	212	195–212	5,479	UBFAU, C	Vernon area	M
(C-8-6)21cab-1	1	343	303–340	5,632	UBFAU, U	Vernon area	GC
(C-9-4)9dab-1	7	462	382–462	5,545	LBFAU, ?	Southeastern Rush Valley	GC
(C-9-4)14bad-1	8	580	580–open end	5,698	LBFAU, ?	Southeastern Rush Valley	GC
(C-9-5)5bbc-1	3	600	uncased below 493	5,595	UBFAU, C	Vernon area	M, GC
(C-9-5)6aab-1	—	—	—	5,585	UBFAU, C	Vernon area	M
(C-9-5)6aab-2	2	170	—	5,597	UBFAU, C	Vernon area	GC

Table A1–3.   Unadjusted and adjusted radiocarbon ages for premodern groundwater samples collected for carbon-14 and stable carbon isotope ratios in and around Rush Valley, Tooele County, Utah.

[Location: See "numbering system" at beginning of report for an explanation of the numbering system used for hydrologic-data sites in Utah See figure 15 for the location of sites sampled as part of this study and table A1–2 for physical information about the sampled well or spring.]

Location	Site ID	Calibrated radiocarbon ages, in years before present, rounded[1]			
		Unadjusted	Tamers (1975)	Ingerson and Pearson (1964)	Fontes and Garnier (1979)
(C-4-5)8cbb-1	25	20,000	18,000	11,000	11,000
(C-4-5)13bad-1	24	17,000	13,000	11,000	11,000
(C-4-5)30aac-2	19	26,000	25,000	19,000	19,000
(C-5-5)3bcc-1	23	10,000	6,200	1,700	1,600
(C-5-5)8dab-1	18	25,000	24,000	14,000	13,000
(C-7-2)29dbc-1	13	17,000	13,000	7,600	7,500
(C-7-3)30acc-1	11	20,000	18,000	12,000	11,000
(C-8-3)3cbd-1	12	32,000	32,000	23,000	22,000
(C-8-4)22aad-1	9	22,000	20,000	15,000	15,000
(C-8-4)36aab-1	10	46,000	45,000	42,000	42,000
(C-8-5) 6ccd-1	5	33,000	33,000	30,000	30,000
(C-8-5)17ccc-1	4	20,000	17,000	13,000	13,000
(C-9-4)14bad-1	8	21,000	19,000	15,000	15,000
(C-9-4)14bad-1	8[2]	21,000	19,000	16,000	15,000

[1] Ages converted to calendar years before present using the Fairbanks 0107 calibration curve (Fairbanks and others, 2005)

[2] Replicate sample included to ensure repeatability of interpreted values

Appendix 2. Groundwater-Budget Uncertainty

The groundwater budget values in table 4 are estimates based on models, assumptions, correlations, or regressions that are fundamentally derived from representative measurements often made at only a few points in time. As a result, these estimates have an associated uncertainty that is difficult to quantify but important to acknowledge. An attempt has been made in the current study to quantify these uncertainties. Each of the groundwater-budget components in table 4, along with the total recharge and discharge, are presented with an uncertainty value (expressed as a percentage of the component value) that is intended to convey the possible range that the actual value might vary. Often, budget components are derived from several variables, and the uncertainty reported in table 4 is that of the variable with the largest contribution to the total uncertainty. The list that follows briefly explains how each of the uncertainties was derived.

Recharge Components

(1) The uncertainty in the *infiltration of precipitation* is based on a sensitivity analysis of the Basin Characterization Model in-place recharge by the authors of the model and documented in Flint and others (2011).

(2) The uncertainty in the *infiltration of unconsumed irrigation water* is based on the reported range of infiltration of unconsumed irrigation water (10 to 50 percent) for similar climatic and hydrologic settings (Feltis, 1967; Clark and Appel, 1985; Stolp, 1994; and Susong, 1995) recognizing that a part of the associated uncertainty comes from estimating ungaged streamflow that is captured and subsequently delivered for irrigation.

Discharge Components

(1) The uncertainty in the *evapotranspiration of groundwater* is based on an error analysis that assumes that the range of evapotranspiration rates are either (i) the high and low measured rates for different vegetation types reported in recent literature (Nichols, 2000; Berger and others, 2001; Reiner and others, 2002; Cooper and others, 2006; and Moreo, 2007) or (ii) +/- 1 standard deviation for rates reported as consumptive use of irrigated crops (Utah State University, 1994). This analysis also takes into account that the uncertainty in precipitation that is subtracted from total evapotranspiration to obtain the estimate of groundwater evapotranspiration is +/- 15 percent (Jeton and others, 2006).

(2) The uncertainty in the *discharge to mountain springs and stream baseflow* is the root mean square error of the regression equation used to estimate streamflow for ungaged streams (Wilkowske and others, 2008; table 4).

(3) The uncertainty in the *well discharge* is the average coefficient of variation (standard deviation divided by the mean) from 15 wells in north-central Utah where at least 5 discharge measurements were made and correlated with electric power consumption records as part of an internal U.S. Geological Survey study (unpublished) to quantify the uncertainty associated with estimating annual groundwater withdrawal from irrigation wells in Utah.

(4) The uncertainty in the *subsurface discharge to Cedar and Tooele Valleys* is assumed to be mostly due to uncertainty in the transmissivity values used in these calculations. Therefore, the reported uncertainty is the coefficient of variation (standard deviation divided by the mean) for the 16 basin-fill transmissivity values listed in table 3.

Lastly, the uncertainty of the total recharge and total discharge was calculated as a weighted average of the individual component uncertainties.